HOW TO MAKE PRINTED CIRCUIT BOARDS

JOEL GOLDBERG, Ph.D.

Professor, Electro-Technology
Macomb County Community College
Warren, Michigan

McGraw-Hill Book Company
Gregg Division

New York	Düsseldorf	Panama
Atlanta	Johannesburg	Paris
Dallas	London	São Paulo
St. Louis	Madrid	Singapore
San Francisco	Mexico	Sydney
Auckland	Montreal	Tokyo
Bogotá	New Delhi	Toronto

Library of Congress Cataloging in Publication Data

Goldberg, Joel,
  How to make printed circuit boards.

  (Electro-skills)
  Includes index.
  1. Printed circuits. I. Title. II. Series.
TK7868.P7G64   621.381'74   79-18137
ISBN   0-07-023634-8

How to Make Printed Circuit Boards

Copyright © 1980 by McGraw-Hill, Inc. All rights reserved. Printed in the United States of America. No part of this publication may be reproduced, stored in a retrieval system, or transmitted, in any form or by any means, electronic, mechanical, photocopying, recording, or otherwise, without the prior written permission of the publisher.

   4 5 6 7 8 9 0   DODO   8 8 7 6 5

Sponsoring Editors: Gordon Rockmaker and Mark Haas
Editing Supervisor: Karen Sekiguchi
Designer: Tracy Glasner
Art Supervisor: George T. Resch
Production Supervisors: Regina R. Malone and Priscilla Taguer
Cover Photography: Martin Bough/Studios Inc.

Contents

Preface v

Chapter 1 Introduction 1

Chapter 2 The Circuit Board 8

Chapter 3 Artwork Preparation 14

Chapter 4 Direct Layout Procedures 39

Chapter 5 Photo Layout Procedures 49

Chapter 6 Silk-Screen Procedures 65

Chapter 7 Etching the Board 74

Chapter 8 Final Processing 82

Chapter 9 Project Description and Construction 91

Index 115

Preface

How to Make Printed Circuit Boards is a comprehensive easy-to-read guide for the practical-minded person interested in producing printed circuit boards. Whether you are a student starting the study of electronics, a hobbyist interested in building electronic projects, or a professional called upon to develop a new circuit board, this book will help you determine the most practical method for your needs. You will start with the basics in design and construction, and become familiar with boards readily available on the market. Different methods of layout are explained, followed by thorough discussions of the ways to produce the actual board, including direct, photographic, and screen processes.

A project for a regulated variable dc power supply is included to illustrate the layout and development of a printed circuit board. The supply voltage is variable from 1.5 to 25 volts dc. It will develop a maximum of 1.2 amperes of current for use in any electronic project. By constructing this project, you will gain hands-on experience in preparing a circuit board as well as a useful addition to your workbench.

Joel Goldberg, Ph.D.

1
Introduction

WHAT IS A PRINTED CIRCUIT?

A printed circuit is an electronic circuit mounted on a base material. The circuit is made of copper foil so thin that it needs a base to support it. The base is also a mounting device, used to fasten the complete package to its case. The type and shape of the actual electronic circuit are limited only by the imagination of the person designing the board.

The name printed circuit arose because the electronic circuit appears to be printed on the base material. In ordinary printing ink is deposited onto paper. The electronic printed circuit gives this same appearance although the circuit is actually a thin layer of copper. The shape of the copper is determined by the layout, or artwork, required for the actual circuit. The final shape is developed by etching, that is, chemically removing some copper from the surface of a blank board. The remaining copper and the base material form the completed printed-circuit board. Such a board is shown in Fig. 1-1. It has been etched, and the mounting holes for components have been drilled or punched, but parts are not yet mounted.

DEVELOPMENT OF THE PRINTED-CIRCUIT BOARD

The printed-circuit board was developed by the electronics industry so that mass-production techniques could be applied to electronic assemblies. Using printed-circuit boards gives a high rate of reliability in production. All circuits are uniform in layout, eliminating the wiring errors common to hand-wired electronic circuits. In production, component parts are often inserted by machine into the finished board. Hundreds of similar circuits can be manufactured this way.

This process is used by the electronic hobbyist for many of the same reasons although hobbyists are not interested in producing large quantities of the same circuit. Many excellent construction articles published each year by a variety of electronic hobby magazines cover

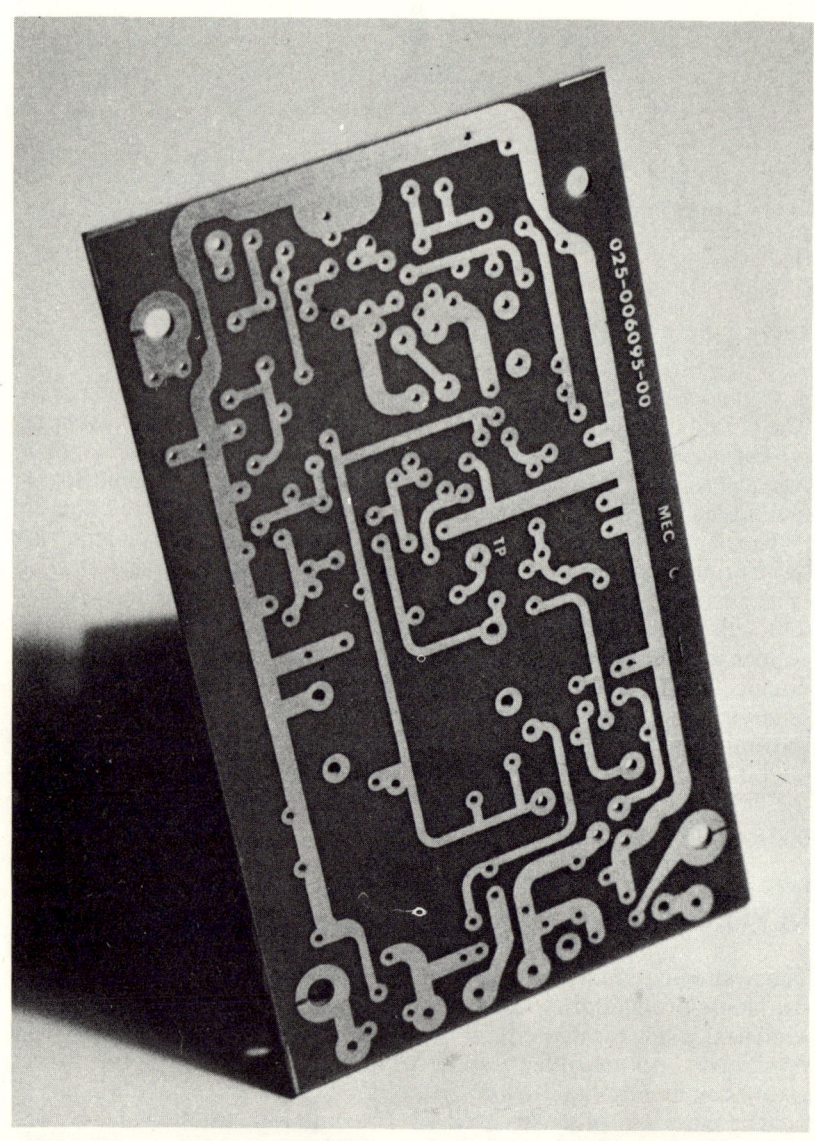

Fig. 1-1 A completed circuit board. This board is ready to have assorted electronic components mounted on it.

amateur radio, computers, television, and similar subjects. Articles often include the artwork for a printed-circuit board. The hobbyist can use this artwork to produce a board or boards in constructing the project. As in industry, using prepared artwork for the printed-circuit board eliminates many errors.

An etched, or printed, circuit consists of a thin layer of copper foil. The final circuit is shaped by etching the copper in a chemical. The copper foil acts as the wire, or conductor, in the circuit. Component parts like resistors, transistors, and capacitors are soldered to the conductive foil to complete the electrical path and circuit.

BOARD CONSTRUCTION

The copper foil used in printed circuits is usually 0.0014 to 0.0042 inch (in) thick. Its support, called the *substrate* or *base,* is either a phenolic or fiber-glass material. Board construction is illustrated in Fig. 1-2. Boards are produced by laminating or gluing a sheet of the copper foil onto the substrate. You can buy boards with copper foil on one or both sides. Whether you use a single or double board depends upon the specific circuit requirements.

BOARD PREPARATION

Before etching, a *resist* material is put on top of the copper foil. The resist protects the copper foil so that it will not be removed when the circuit board is etched. Once the electronic circuit has been designed, it is copied onto the copper foil in one of several ways. All the foil *except* where the resist is placed is removed from the board by etching. The

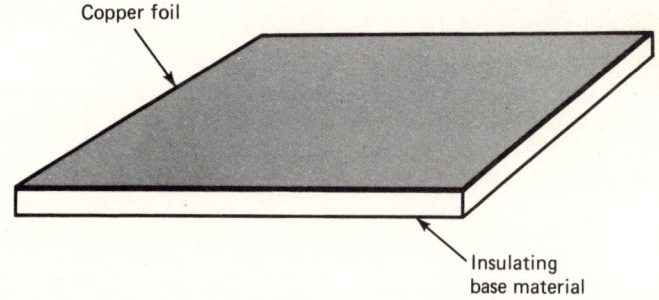

Fig. 1-2 **The circuit board consists of a thin layer of copper foil and a supporting base, or substrate.**

etching solution, or *etchant,* attacks the copper foil and removes it from the board. The end result is the electronic circuit attached to the substrate material. The board is removed from the etching solution, cleaned, and inspected.

After inspection, holes are drilled in the board to be used for mounting parts, like resistors, transistors, and capacitors. The parts are mounted and soldered to the board. Wires are added to connect parts mounted off of the board to those on the board. The printed circuit board is now ready for use.

At first some of the very complex printed-circuit boards in use today may seem overwhelming, but the boards are not difficult to produce if the basic steps shown in Fig. 1-3 are followed. You start with an electronic schematic diagram of the circuit. This diagram is the actual wiring layout of the circuit in a shorthand form using standard

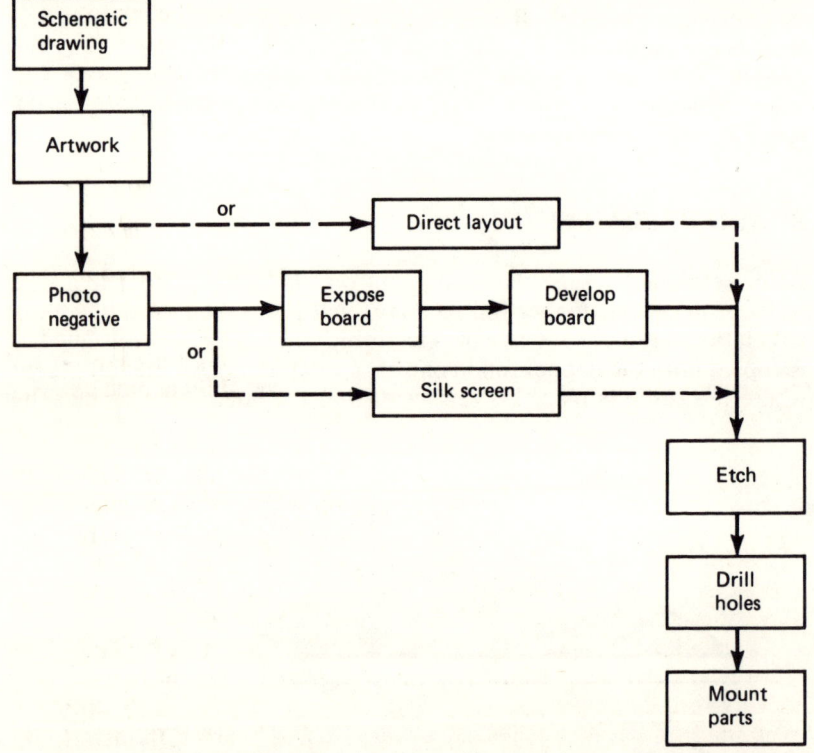

Fig. 1-3 The steps involved in development of a printed-circuit board. Direct, photographic, or silk-screen methods are used to prepare the board for etching.

electronic symbols. The schematic diagram shows how the circuit is wired in a graphic format.

The next step is to translate the schematic diagram into the *artwork* used to make the printed-circuit board. This is probably the most challenging part of the whole process. Artwork may seem difficult at first, but once some simple rules are understood, it becomes both interesting and satisfying.

There are two methods available for the third step: (1) lay out the artwork directly on the printed-circuit-board material or (2) lay out the artwork on a piece of photographic film. This decision depends upon the person making the board and such factors as the complexity of the circuit, the number of identical boards to be made, the materials available, and whether someone else has already laid out the board.

Once this decision has been made, the artwork can be reproduced on the circuit-board material. One method requires artwork to be transferred to a piece of photographic film. The film is developed and placed on top of a specially prepared circuit board. Exposing the photographic negative and sensitized board to light transfers the artwork to the circuit board. In the other method the artwork is redrawn directly on the copper board material with an ink or paint resist. This second method is excellent for production of one-of-a-kind boards.

HOW TO USE THIS BOOK

The processes described so briefly in this chapter are covered in detail in the chapters that follow. Each step is clearly defined to help the reader learn how to make printed-circuit boards. This book tells how to make the boards by several processes, all of which are easy for the hobbyist. Many of these processes can also be used by people working in the field of electronics.

Although this book does not aim to show how to build specific projects, one project is used to illustrate how to make printed-circuit boards. A regulated variable dc power supply is used as an example of how to lay out, develop, and produce electronic printed-circuit boards.

Each of the following chapters is devoted to one phase of the production of circuits. A flowchart, similar to Fig. 1-3, is given as Fig. 1-4. Chapter numbers are included to assist the reader.

The book is arranged so that the reader can start at its beginning and learn how to develop artwork required to make a circuit board. After the artwork is complete, several options are open. Each option is presented so that the reader can select one and follow it through to the completion of the project.

The options are offered because everyone has likes and dislikes. We all know the limitations of our pocketbooks, equipment, and

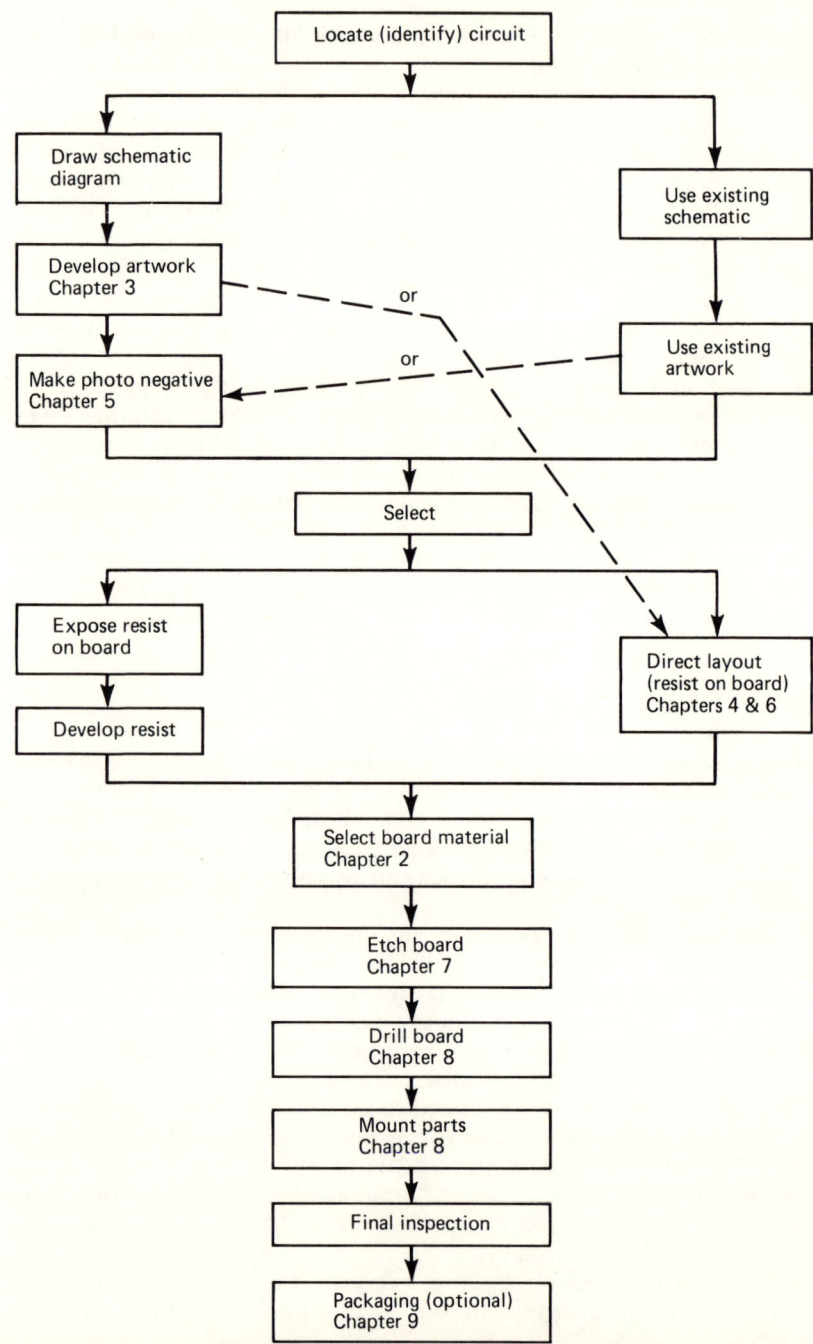

Fig. 1-4 A flowchart showing the steps involved in developing a circuit board. Each major step is related to a chapter of this book.

capabilities. Select the process with which you feel comfortable and follow it through until the project is complete. Success should be yours if you take your time and consider each step before starting it.

Laying out and etching printed-circuit boards is not difficult. The pride of accomplishment is well worth the effort involved. Try several different processes as you spend time making boards. Enjoy yourself. This kind of work is fun. The satisfaction accompanying success is very rewarding.

2
The Circuit Board

The printed circuit starts out as a blank. Manufacturers of materials used to produce the final printed circuit usually are not concerned with how the board will look when it is complete but with the basic materials. Most blank circuit boards consist of a substrate and copper foil. The type of substrate, the amount of copper, and how the copper is fastened on are all important.

Hobbyists normally do not need the wide range of circuit-board materials available to industry and government. Hobbyists use small quantities of board materials, which are often available only from stores specializing in such sales. Unusual types of board materials are not readily available to the public. Anyone requiring special sizes or materials should be in touch with manufacturers or their representatives about buying these products.

BOARD CLASSIFICATION

Circuit boards are classified as either *flexible* or *rigid*. This refers to the base material. The copper, being so thin, will normally bend. Most blank boards used by the hobbyist start out as rigid boards. This does not imply that the board cannot flex but that flexing is kept to a minimum. The board cannot be bent like soft plastic or paper.

Flexible boards are made of a sandwich of copper foil laminated between two pieces of soft plastic. The plastic permits bending and shaping the board. Since special equipment must be used for flexible boards, they are not normally made by the hobbyist.

Rigid boards are produced as shown in Fig. 2-1. Copper foil is laminated, or glued, onto the surface of the base material. Boards are produced with copper foil on either or both sides of the substrate to make single- or double-sided boards. Most simple printed circuits use a single-sided board. Double-sided boards are used for more advanced designs. Techniques for making both kinds are similar. The selection of

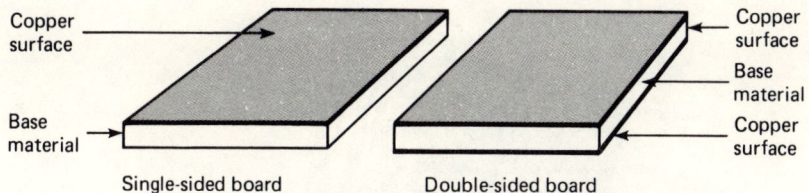

Fig. 2-1 Construction of single- and double-sided circuit-board blanks.

the proper board is up to the user and designer of the circuit. Factors which determine selection are presented later in this book.

Figure 2-2 shows a cross-section of a circuit board. The thicker portion of the board is the base, or substrate. The copper foil appears as a thin dark line at the edge of the board. The picture shows the copper-foil side of a board before any etching, or copper removal, has been done.

The circuit board shown in Fig. 2-3 is a completed board. The circuit was laid out on it and it was then placed in an etchant solution, which chemically removed all unwanted copper. The end result is a copper-foil electronic circuit attached to a supporting base.

Double-sided boards look more like a covered sandwich. Copper conductive foil is glued onto both sides of the substrate. The circuits are then etched on each piece of copper. This forms conductive paths on both the top and the bottom foil. Connections are made from one side to the other by eyelets or rivets. Using a double-sided board often results in a smaller circuit board overall. It also permits more complicated circuits to be used. Alignment of the two circuits before etching is critical when making a double-sided board.

THE COPPER FOIL

There are several subclassifications of circuit boards. One relates to the copper foil used on the board. Most of the conductors used in printed-circuit boards are made of copper foil. The foil is manufactured by rolling copper bars between steel rollers until it is very thin or by electroplating copper onto a stainless-steel drum. Most of the foil used today is made by the electroplating process. Once the copper foil

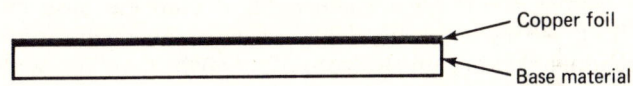

Fig. 2-2 In a circuit-board blank the copper foil is glued onto the thicker base material.

Fig. 2-3 An etched circuit board. This board has had the copper foil etched in order to produce an electronic circuit pattern.

is plated onto the drum, it is stripped off in large sheets. These sheets are fastened to the base material with a bonding material that must be able to withstand heat, chemicals, and stress in order to keep the package intact.

Copper foil is manufactured in several thicknesses from 0.0014 to 0.0042 in. Boards are classified by the weight of copper in each square foot of surface. Typical ratings range from ½ to 3 ounces per square foot (oz/ft^2). The thickness of the copper is directly related to its weight per square foot.

The weight ratings are also used to indicate the amount of electric current the foil will carry. (Table 2-1). Two dimensions are given in this table, the thickness of the copper foil and the width of the conductive path on the board. These two dimensions give a cross-sectional area. The cross section of a conductor determines the maximum electric-current capabilities of the conductor. Additional current flow through the conductor generates heat. The heat may cause the conductive foil to lift by softening the glue which holds it onto the base material. Further heat will melt the conductor, causing an open circuit. Still further heat may set the substrate on fire. Obviously all these deteriorations are to be avoided.

Selection of the copper weight is not always possible for the hobbyist. Use of adequate conductor paths will help ensure a safe

Table 2-1 Thickness of Copper and Width of Conductor Paths for Circuit-Board Design

Weight, oz/ft^2	Foil thickness, in	Conductor width, in	Current rating, A
½	0.0007	0.005	0.13
		0.010	0.50
		0.020	0.70
		0.030	1.00
1	0.0014	0.005	0.50
		0.010	0.80
		0.020	1.40
		0.030	1.90
2	0.0028	0.005	0.70
		0.010	1.40
		0.020	2.20
		0.030	3.00
3	0.0042	0.005	1.00
		0.010	1.90
		0.020	3.00
		0.030	4.00

board design. The values given in the table are minimums. Using widths greater than those shown ensures that a board will not fail because of overcurrent problems.

THE BASE MATERIAL

Another subclassification for a circuit board is its base material. This substrate is an insulator. Its sole purpose is to support the thin electronic circuit etched in the copper foil. Base materials must meet several criteria, including mechanical strength, operating-frequency limitations, and machinability.

The base material consists of some insulating material which is combined with a plastic under heat and pressure. The result is a strong supporting base for the copper foil. Most circuit boards used by the hobbyist have either a phenolic or a fiber-glass base.

Phenolic boards are impregnated with a resin material for strength and are available with various ratios of resin to the phenolic. These boards are rated by the amount of resin (Table 2-2). Boards containing a small percentage of resin are rated with one X. As the percentage of resin increases, so do the number of X's used. Most boards available to

Table 2-2 Comparison of Properties Important in Selecting Board and Base Materials, Ranked from 1 (Low) to 10 (High)

Board type	Temperature resistance	High frequency	Base material
X	3	1	Paper-base phenolic
XX	2	2	Paper-base phenolic
XXX	2	3	Paper-base phenolic
XP	3	1	Paper-base phenolic
XXP	2	2	Paper-base phenolic
XXXP	2	3	Paper-base phenolic
G10	6	6	Fiber-glass-base epoxy

the hobbyist have either an XX or an XXX rating. If the board is designed to have holes punched in it by machine, it also carries a P rating. Ratings of XXP and XXXP are also used for boards with a phenolic-resin-base material.

A second common base material is fiber-glass resin. Circuit boards using a fiber-glass-resin base are rated G10 by the industry. Fiber glass is used for several reasons. Boards made of fiber glass do not break as easily as those made from phenolic resins. They can be bent slightly without cracking the substrate material and will withstand shock better than the other boards. If the electronic device containing the circuit board is dropped, the board may break, especially when component parts are mounted on the board.

Boards also have a tendency to change physical dimensions when exposed to extremes of temperature. Most parts connected to the copper foil are held in place with solder. Solder is not normally a good physical bonding agent. It makes better electric connections. Changes in board size cause the solder connections to break, leading to circuit failure.

Table 2-2 also gives frequency ranges for the different base materials. Boards made of phenolic resin should not be used for frequencies above 10 megahertz (MHz). The base material will not handle the higher frequencies without breaking down. This will cause at least a loss of signal and possibly a ruined board.

Fiber-glass-resin boards can handle frequencies up to 40 MHz without failure. If you are making boards for use in transmitters or receivers, G10 boards should be used. They are easily identified by the green base. Phenolic-resin boards are normally brown.

One other item to consider when selecting a blank circuit board is its cost. Table 2-3 shows the relation of board base material and weight of copper to the cost. The single-sided XX board is given a rating of 1.0 and used as a reference base. Cost indexes are given for each of the various types of boards. The index price shows the relative price of

Table 2-3 Cost Factors for Single- and Double-Sided Boards (Single-Sided 1-oz-Foil Board = 1)

Type	Foil weight, oz	Nonsensitized		Sensitized	
		One side	Two sides	One side	Two sides
XXXP					
	1	1.00	1.40	1.76	2.48
	2	1.10	1.60	1.84	2.68
G10					
	1	2.52	2.64	3.48	4.20
	2	2.64	2.68	3.60	4.28

other boards. For example, a G10 one-sided board has an index rating of 2.52, which means it costs 2.52 times the cost of the XX reference board. The last two columns give the index for boards with a light-sensitive resist coating. These boards are more expensive because of the extra handling and processing involved.

OVERALL CONSIDERATIONS

Selecting the proper material for making a printed-circuit board includes many factors. The most important relate to cost, copper thickness, base material, and whether the board is single- or double-sided. Fortunately for the hobbyist, the decisions are limited to a few major points. Most boards available are going to be either XXXP or G10 material. The copper weight will probably be 1 oz. Selection is then dictated by frequency applications, circuit design, and availability of the board material locally.

3

Artwork Preparation

The purpose of artwork preparation is to develop a layout for the final circuit board. The artwork is used to prepare the foil on the board for etching. It is easier and less expensive to do this layout on a sheet of paper than on the copper foil. Mistakes are easy to correct with an eraser. There is no need to paint over or remove any artwork from the board.

There are several things to consider when making up the artwork. If artwork is already available, there is no need to do it over. Move ahead a few chapters in this book and read how to proceed after the artwork is done. However, if there is a need to do your own artwork, read on.

SCHEMATIC DIAGRAM

The first thing required when making a layout for a printed circuit board is a good schematic diagram of the circuit you plan to make. The circuit shown in Fig. 3-1 of a variable regulated dc power supply will be used to illustrate the steps involved in preparing artwork. This supply is capable of handling a maximum of 1.2 amperes (A) of current at up to 24 volts (V). The output varies from 1.5 to 24 Vs, depending on the setting of the variable voltage control. This is a functional schematic. The circuit works. This power supply is excellent for general repair work or as a replacement for batteries when using a portable electronic device in the home.

Look at the schematic diagram. The input is on the left side, and the circuit output on the right. The most positive voltage lines are at the top and the circuit common is at the bottom of the drawing. This placement is almost always used in schematic diagrams. Slight modification of this placement is used when laying out the artwork for the circuit board.

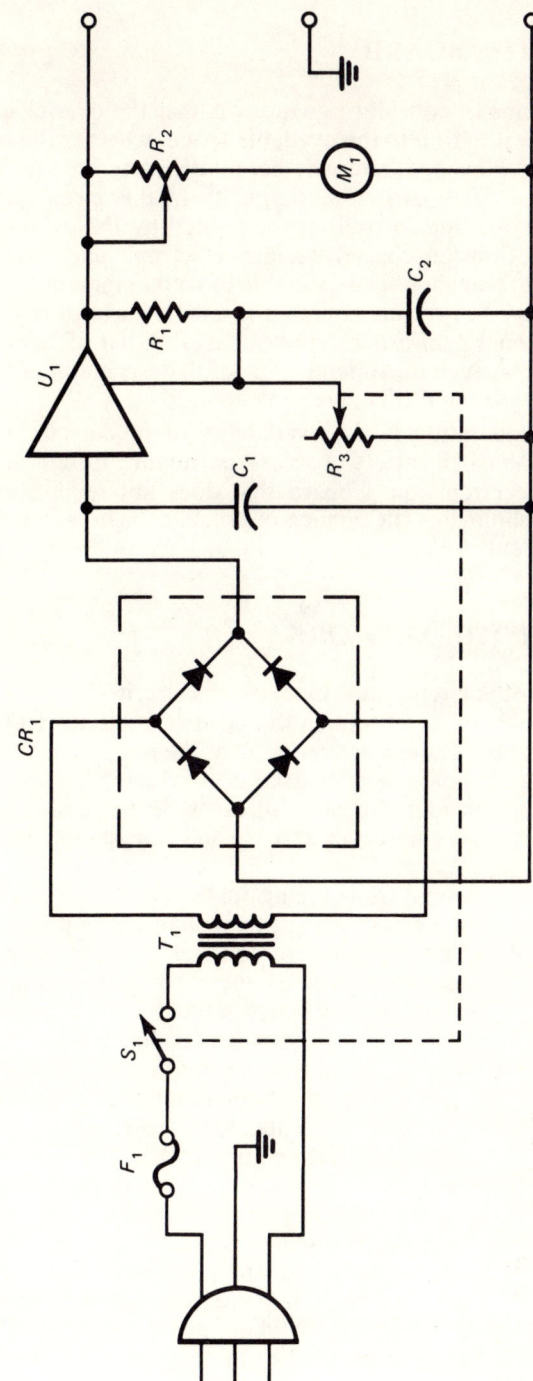

Fig. 3-1 Schematic diagram of the variable regulated power supply used as an example in this book.

CHOOSING THE BOARD

Some of the items to consider now are whether the overall size of the board will allow it to fit into the available space, whether the board will be single- or double-sided, and whether all the parts used in the circuit are mounted on the board. The simplicity of this circuit calls for a single-sided board. The overall size is limited by the cabinet. In this case, the cabinet was purchased at a local electronic parts store (part of a national chain) and should be available in your community. The parts and cabinet (Fig. 3-2) are all available at a local parts store or through one of the national mail-order supply houses. A list of some of these supply sources is given in Appendix C. A full description of the power supply and how to build it is given in Chap. 9.

Another consideration is the availability of precut circuit boards. These are available in a variety of sizes from about 2 by 4 in. to 12 by 12 in. Lay out the circuit on a board that does not require cutting, if possible. This eliminates the chance of cracking or breaking the board while trying to cut it down.

LAYING OUT THE ARTWORK

The artwork for the circuit must meet two basic criteria: (1) it serves as a master from which to reproduce the electronic circuit on the board and (2) it must allow enough space for all of the parts to be mounted on the board. Several approaches are used in developing the final artwork, but none can be considered best in all cases. Each person selects the approach that is most comfortable to use and suitable for the available materials.

A point to keep in mind while laying out the board artwork is spacing between lines or conductors. An arc-over may occur if two lines are too close together. Spacing between conductive paths depends upon their voltage difference. As the voltage increases, the spacing must also increase (Table 3-1). The spacing starts at 0.025 in for a range of up to 150 V dc or peak ac. If the voltage between conductors is as high as 500 Vs, the spacing must be 0.100 in. Spacing for voltages higher than 500 V is figured at 0.002 in/V. These are minimum values. There is no rule that says the spacing cannot be greater than these values.

The size of the artwork will depend upon whether a copy camera is available. The camera is used to reduce the artwork to its final size. In commercial circuit-board development the artwork is often made twice or four times larger than the true size (written 2× and 4×). Working on such a large scale makes it easier for the person laying out the board. Any minor errors due to line alignment are reduced by a factor of 2 or 4 when the final board is completed. Many people working in this field use a size that is twice as large (2×) as the final product. This is a

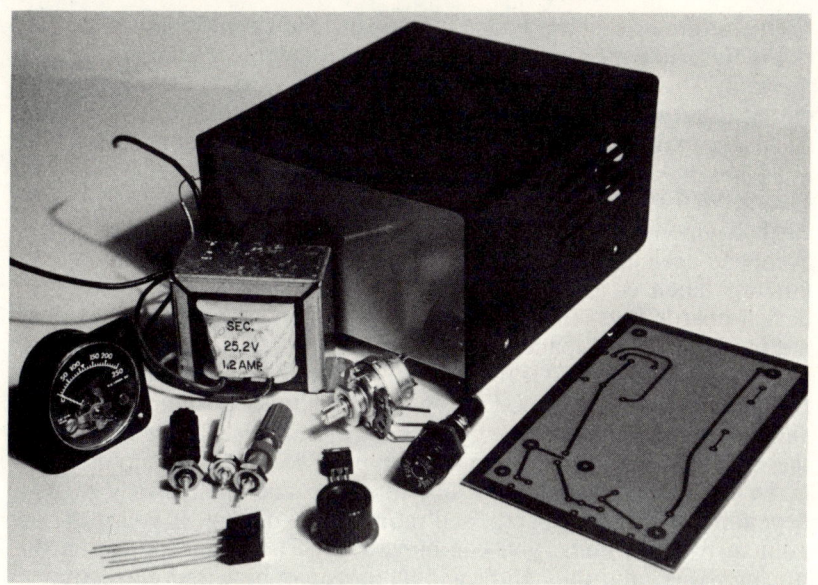

Fig. 3-2 Component parts and cabinet used for the variable dc power supply.

compromise that keeps the enlarged original artwork to a reasonable working size. Persons who have no copy camera work on a ratio of 1:1, or 100 percent of the finished size. This is slightly harder to work with, but it is the most practical under these circumstances. Size relationships are shown in Fig. 3-3.

One of the most convenient ways of laying out the circuit board is to use a grid paper. It is available with ruled lines spaced at 0.05, 0.10, and 0.125 in. Most electronic circuit boards use a 0.10-in grid. Lines run both horizontally and vertically on the page. The results is a page full of small boxes. The ink used to print the grid lines is usually light blue or black. The grids are used as a guide in laying out the final artwork. Sample grid papers are shown in Fig. 3-4.

Table 3-1 Spacing Between Conductor Paths as It Relates to Voltage Differentials between Paths

Voltage between conductors (DC or AC peak)	Minimum spacing, in
0–150	0.025
151–300	0.050
301–500	0.100
Over 500	0.002 per volt

One of the easiest methods of laying out the circuit-board pattern is to use the grid paper as an underlay. Place a sheet of tracing paper over the grid. This paper is used to make a rough working drawing of the circuit. First mark the tracing paper with an outline of the circuit board to ensure that the artwork and the board end up the same size.

Layout Methods
Each component mounted on the board must fit in its proper space. Someone designing a circuit board may laugh at the above statement, but this laugh is usually premature. How often do those of us who design boards find out *after* the board is complete that one part does not fit in its allotted space! One of the best ways to reduce this problem to zero (or almost zero) is to use a set of models for each part that is to fit on the board. These models, or *dolls,* can be cut from cardboard as shown in Fig. 3-5. Their size will depend upon the scale used to lay out the board. A set of dimensions for typical electronic components is given in Table 3-2. Note that transistor and capacitor size vary greatly, depending upon the capacity and rating of the device. If you make up your own set of dolls, you can move the parts around on top of the tracing paper until they all fit and all parts can be wired correctly.

To lay out your own sets of dolls use the guidelines shown in Table 3.2. Dimensions are given in this table for many of the readily available components used on circuit boards. The dimensions are given in tenths of an inch (0.1 in). If you are using graph paper laid out with 10 squares per in each square is equal to 0.1 in. Count the number of squares to determine overall spacing between component leads. Be sure to allow sufficient space for the rest of the components, areas which extend beyond the leads of the part.

Keep in mind while laying out a circuit board that the parts mount on the back side of the board. These parts will be on the base side of a

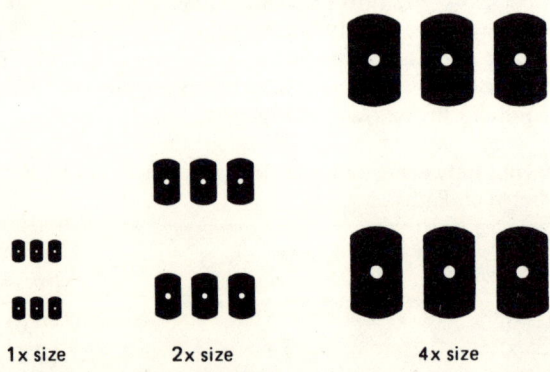

Fig. 3-3 Circuit board artwork is either 1×, 2×, or 4×. Mat is actual size, twice as large, or four times as large. (*Bishop Graphics, Inc.*)

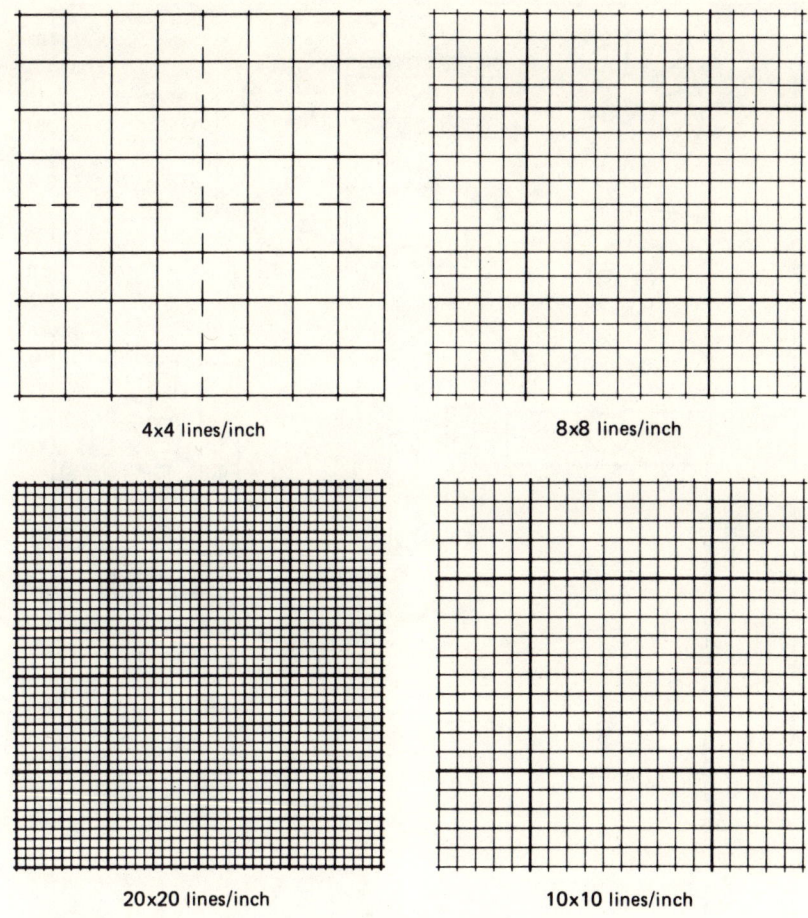

Fig. 3-4 **Grid papers commonly used for layout of circuit boards.**

single-sided board. Certain parts such as transistors and ICs have their leads arranged in a definite order, and it is easy to overlook this backward viewing of the parts when designing a board. If you forget to check these pin connections, you could end up with an unworkable layout.

Another approach to this same problem is the use of commercial outlines, available in 2× and 4× sizes. One source of supply is Bishop Graphics, listed in Appendix C. The dolls are shown in Fig. 3-6 and are reusable. They function like the dolls described above.

Another excellent method is to use drafting templates (Fig. 3-7), available from several drafting-supply manufacturers. The template can be used instead of the dolls or puppets when laying out the circuit

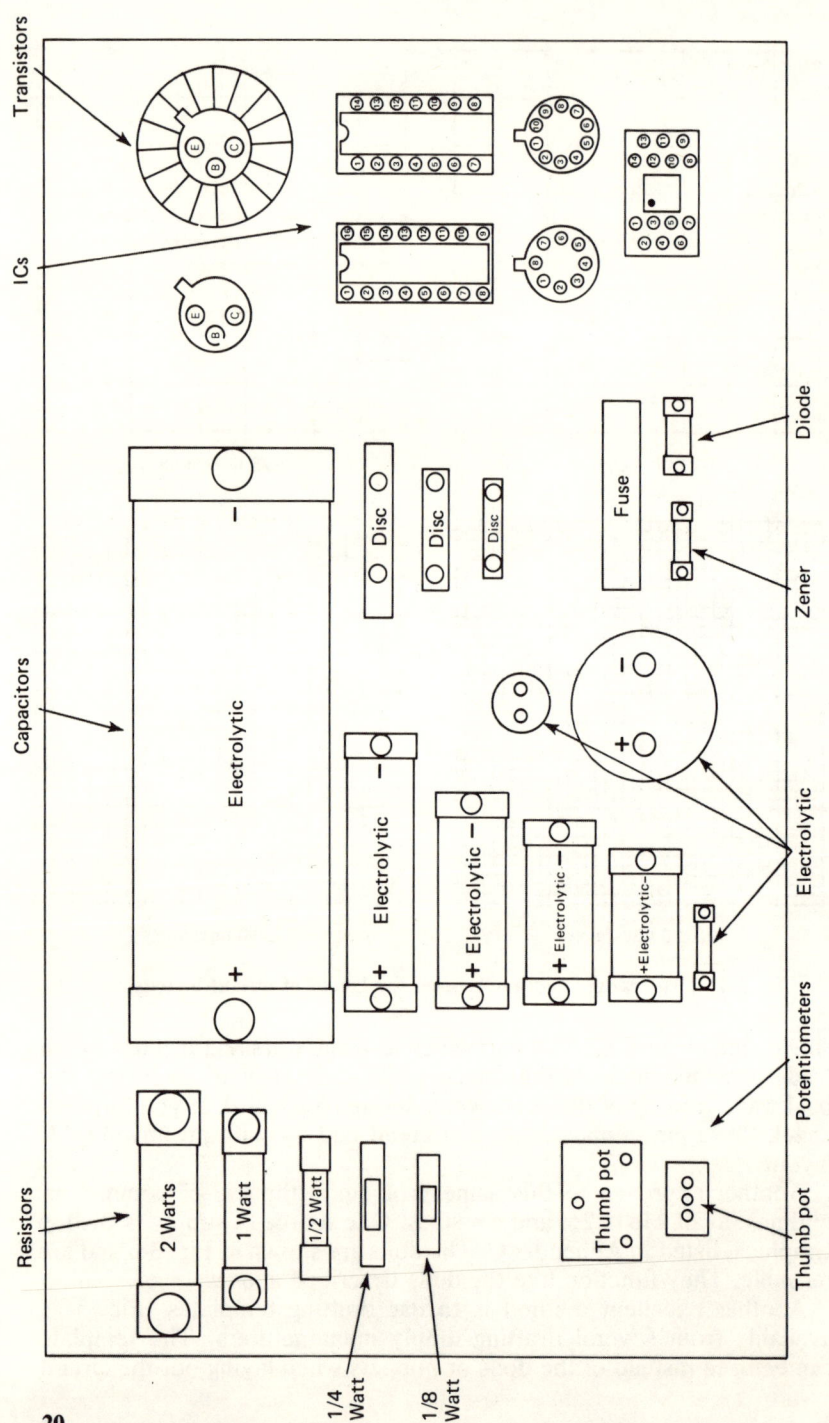

Fig. 3-5 Paper cutouts, called dolls, used to position parts during board-layout attempts.

Table 3-2 Artwork Sizes for Various Electronic Components

Component	Spacing
¼-W resistor, signal diode	0.4 in (4 boxes) between leads
½-W resistor, power rectifier diode	0.5 in (5 boxes) between leads
Disc capacitors	0.3 in (3 boxes) between leads
Radial-lead electrolytic capacitors	0.2 in (2 boxes) between leads
Other resistors, capacitors	Measure for spacing
Transistors (small signal)	0.2 in (2 boxes) between leads
DIP ICs with up to 18 pins	0.1 in (1 box) between pins, 0.3 in (3 boxes) between rows of pins
DIP ICs with more than 18 pins	0.1 in (1 box) between pins, 0.6 in (6 boxes) between rows of pins

board. They are available in 1×, 2×, and 4× dimensions. (See also the discussion of templates under the Tape-and-Dot Process in Chap. 4.)

Once the component parts are positioned and connecting lines drawn, the final artwork is developed on the sheet of tracing paper. Some people prefer to lay another sheet of tracing paper on top of the

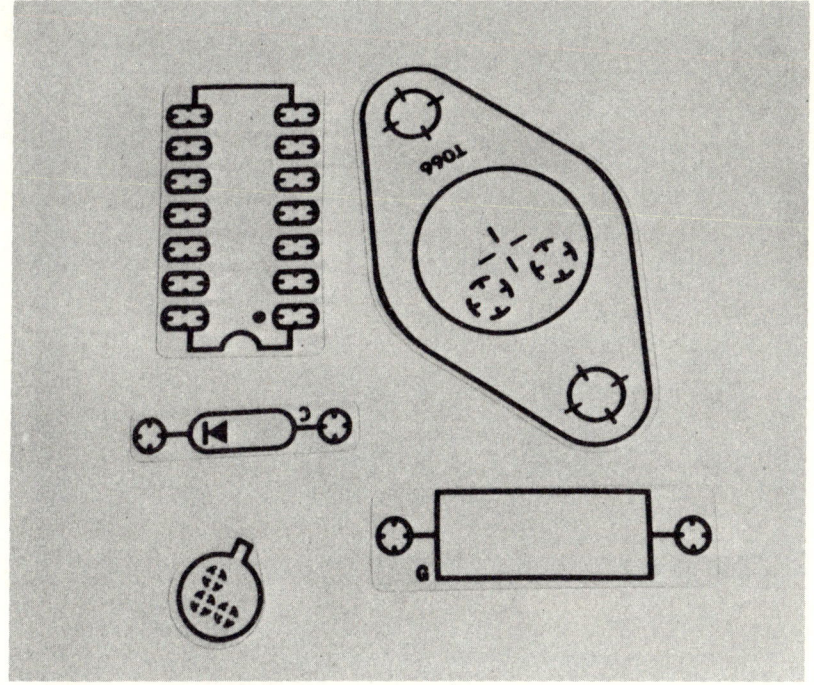

Fig. 3-6 Commercial dolls, or Puppets, used for board layout, available in either 2× or 4× sizes. (*Bishop Graphics, Inc.*)

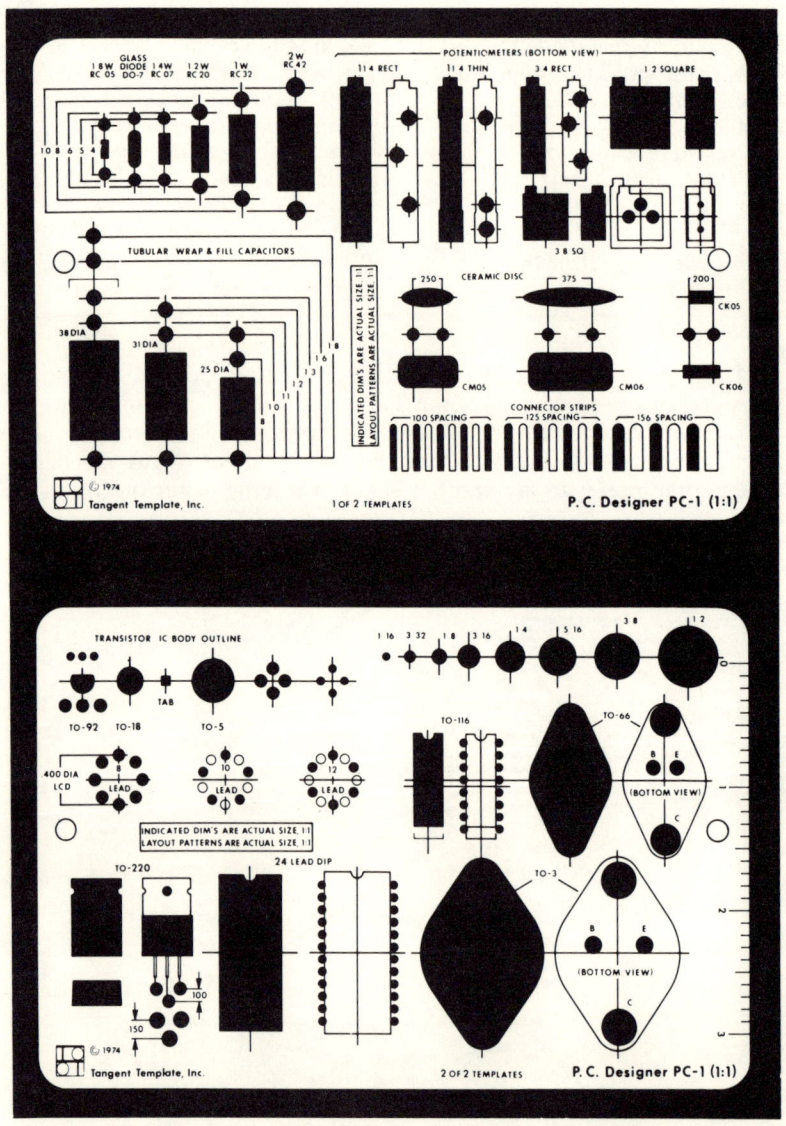

Fig. 3-7 Drafting templates used to lay out circuit boards. These templates are actual size. (*Tangent Template, Inc.*)

one on which the parts are laid out. Each point of contact for a part is marked. Lines representing wires are drawn on this sheet rather than the one with the parts on it. Using a second sheet of paper allows the designer more leeway when attempting to connect all required points on the layout of the board. The dolls can be repositioned if necessary without doing a lot of erasing or redrawing at this stage.

Guidelines
At this point you are probably saying, "This is all well and good, but aren't there some rules to follow in layout work?" You are absolutely correct. The guidelines for board layout are as follows (see also Fig. 3-8):

1. Have a schematic diagram available for the circuit.

2. Determine which components are *not* to be mounted on the board, for example, transformers, controls, variable capacitors, fuse holders, and switches.

3. Arrange the board so that input connections are on one end and output connections are on the opposite end.

4. Common, or ground, should be routed so that it completely encircles the board.

5. Line direction should be either left to right or up to down. This is not always possible, and exceptions are permitted.

6. Each connection to the board must have its own pad. A *pad* is a small circular piece of foil to which a solder connection is made. Do not attempt to mount two components on the same pad. Move something to make room for additional pads.

7. Lines (or conducting paths) cannot touch each other. It may be necessary to reroute a conducting path if this happens.

8. You may route conducting paths underneath other components. *Do not* route these paths between the leads of a transistor.

9. Jumper wires between pads are allowable. Try to minimize them. Rearranging the artwork will often eliminate the need for a jumper wire.

Some of the do's and don'ts are illustrated in Fig 3-8. Try to follow these rules when laying out the circuit board. This is no time to get carried away with wild artwork. Keep the lines and pads in a practical format. The end result will be a much better board.

Read the above guidelines at least twice. You are now totally confused or too concerned, or have decided to be brave and try to

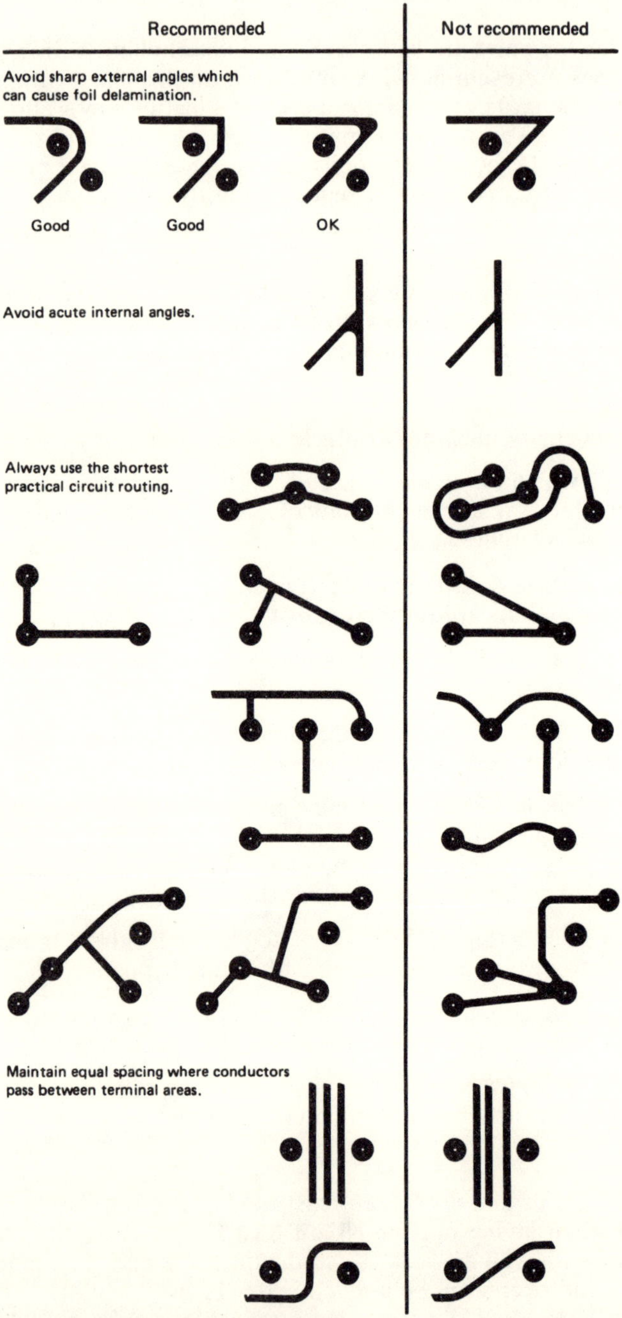

Fig. 3-8 Recommended layout practices for circuit boards. (*Bishop Graphics, Inc.*)

follow them. Keep in mind that they are only guidelines. Successful boards are in use which do not follow *all* the guidelines.

Let us review these guidelines. A second explanation should help eliminate any confusion.

1. Use of a Schematic Diagram   You have to translate the information found on the schematic diagram into marks and lines on a circuit board. Many shorthand symbols are used on the schematic. For instance, common or ground connections are not all shown as solid lines. Often a symbol is used which indicates a common connection. This is not permitted on the circuit board. All common leads have to be connected or the circuit will not function.

The same holds true for other wiring. Often the B+ (positive high-voltage) leads are drawn without showing them connected. On the circuit they are wired together. Input and output leads have to be wired to some sort of connecting terminal. These, too, are different on the board. A good schematic diagram will help you see the total wiring package. It will go a long way toward minimizing circuit-board layout errors.

2. Off-Board Components   Printed-circuit boards are not always self-contained. They often have controls, switches, and other components which are physically mounted to the control panel of the device. Deciding which parts should be mounted off the board is important. This also includes components mounted on their own boards but connected to the main board. A good example is the board on which digital readouts are mounted, which is often separate from the main board in an assembly. Each part mounted off the main board must be marked. All wires going to these off-board components must be located and identified.

3. Isolate Inputs and Outputs   A speaker placed close to a microphone generates an acoustical feedback that sounds like a high-frequency squeal. The same sort of thing happens in an electronic circuit when the input and output connections are too close. The resulting high-frequency whine or squeal may offset any advantage in using the circuit.

Proper procedure in laying out a circuit board is to isolate input and output by placing them at opposite ends of the board. The schematic diagram shows them separated. You should do the same on your board layout.

Most rules have an exception. This is also true with circuit boards. When a board using connections located on one edge is being laid out, the input and output connections land at the same end of the board. When this occurs, separate them as much as possible by putting other wires in the space between them. Figure 3-9*a* shows a standard board template and Fig. 3-9*b* shows a template with single edge connections.

Fig. 3-9 Types of layouts for circuit boards: (*a*) standard (*b*) edge connector boards. (*Tangent Template Co.*)

4. Common Connections Unless you have some unusual type of housing for the circuit, it has to be mounted in a cabinet. The cabinet is usually at ground potential as far as operating voltages are concerned. Mounting devices are often bolts and spacers, which hold the board a set distance from the cabinet. The use of a common, or ground, border around the perimeter of the circuit board helps minimize short circuits for operating voltages or signal circuits if the edge of the board or its mounting bolts should touch the cabinet.

5. Line Direction It is easier to lay out a circuit board when all the conductive paths go in the same direction. Using horizontal and vertical path lines minimizes layout confusion and error. Schematic diagrams use this same approach. The board layout often follows the lines of the schematic. Why fight it? Use the simplest, most direct approach possible when laying out the artwork for the board. Save fancy work for a time when you are more experienced and have had success with several layouts.

Keep in mind that there are times when you must make exceptions. If you have a situation similar to that shown in Fig. 3-10, then do as shown and run a series of diagonal lines between the pads.

6. Individual Pads A pad is a place on the board where an electric connection is to be made. The pad may be used to connect a lead from a component or to connect wires from off-board components to the board. A general rule is that each lead has its own mounting pad. Two leads connected to the same point in the circuit require two pads. Three leads require three pads, and so on. Multiple pads are permitted. Be sure that each component to be connected to the board has enough mounting room. It is very difficult to modify the board or component after the board is made to correct errors due to a lack of foresight.

7. Line Spacing Conductive paths, or lines, cannot touch each other. If this occurs, short circuit results. Review the paths required for conductive lines. Move parts or conductor placement in order to be sure that paths do not touch.

8. Conductor Routing Conductors can be routed under the body of an electronic part in order to complete the path for the circuit. To make such paths it may be necessary to reposition parts on the layout.

Do not run conductive paths between the legs of a transistor. It leads to many complications. You may have to redesign the complete board to avoid this arrangement. The result in circuit operation is   worth the effort.

9. Jumper Wires The alternative is to use jumper wires. There are times when jumper wires connecting two paths or components cannot be avoided. Try to minimize jumpers by routing lines between the legs of other components (see Fig. 3-11).

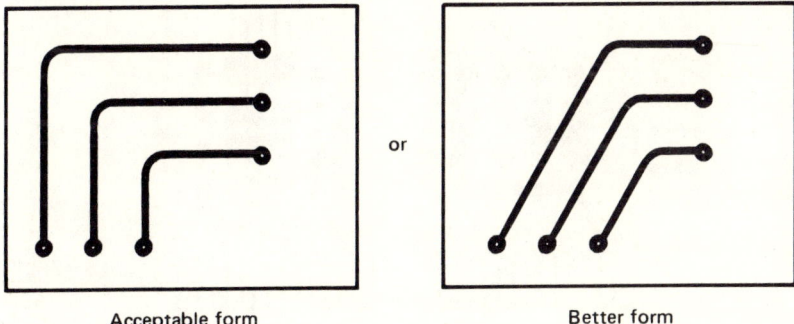

Acceptable form   Better form

Fig. 3-10 Use common sense when laying out the board. Take advantage of the area in the middle to shorten line length.

How the Guidelines Are Applied
Now is a good time to try out these guidelines using the power-supply circuit shown in Fig. 3-1. The first decision to make is to determine the overall size of the unit and the size of the circuit board. The actual board layout can begin once these have been decided. Make a doll for each component part. These will be used to help make both the above decisions.

At this point another decision has to be made: Which parts used for the circuit are to be mounted on the circuit board? That is, which parts are to be mounted off the board? The choice for mounting parts is determined by the circuit requirements. For a power supply certain components *must* be mounted in a way that is not possible on the board. These parts include the output voltmeter, the variable voltage control, the primary lead fuse, and both input and output terminations. Placement of the power transformer is optional. It may be a part of the circuit board or fastened to the cabinet. In this particular layout we decided to make the power transformer a part of the circuit board.

Layout
The procedure for doing the layout follows.

Making the Actual Off-board Components   Start with the identification of off-board components. You may wish to circle these with a colored marker on the schematic diagram, as illustrated in Fig. 3-12. Next, place an x next to each connection on the schematic that goes to each off-board component. Be sure to include an x for each lead of each component. In some cases you may have two x's. This is where two parts are connected together and use wiring on the board for part of the circuit. (There are none of these on the model schematic.)

Place either a letter or a number next to each x. Try to keep them in a

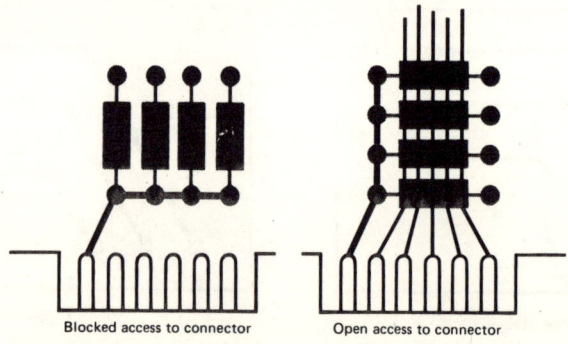

Fig. 3-11 Turning the components 90° allows leads to be run under the components.

sequence to minimize overlooking a connection later. The schematic shown in Fig. 3-13 has off-board parts identified; x's show where leads go off the board and each terminal is marked with a letter. It's a good idea to skip the letter I in order to minimize confusion with the number 1 later.

Parts List Now make up a parts list. If you buy the component parts at this time, check out sizes for circuit-board blanks as well as cabinet dimensions. The unit built as a model used a 3- by 5-in board. The cabinet selected for the project is 5¼ in wide, 3 in high, and 5⅞ in deep.

Parts list

Quantity	Part	Description
1	R_1	220-Ω ½-W resistor
1	R_2	50-Ω trimmer potentiometer
1	R_3	5-kΩ variable control
1	C_1	1000 μF 35-V capacitor
1	C_2	2.2 μF 35-V capacitor
1	T_1	24 to 25-V 1.2-A transformer
1	CR_1	2-A 100-PIV bridge rectifier
1	M_1	0 to 1 mA meter
1	F_1	1-A slow-blow fuse
1	U_1	LM 317 regulator

Assorted wire connectors, line cord, heat sink for U_1, fuse holder, cabinet, and hardware required in addition to the above parts.

Dolls Use the dimensions of the component parts to make up a set of dolls to be used for layout purposes. Each doll is the exact size of the base of the component part. If there is an overhang, or if the overall size of the part is larger than its base, use the overall dimensions when making the doll. This will permit correct part placement and avoid the last-minute discovery that not all of the parts will fit. Include in the dolls any requirements for leads extending beyond the body of the part. The dolls used for this layout are illustrated in Fig. 3-14. Circles shown at the ends or edges of the dolls are allowances for leads. Pads will be placed on the board at these points. Dolls are better to use than the actual components because they are so much easier to handle. Be sure to label each doll with its part name and number before using it. You are now ready to start the actual layout of the board.

Laying Out the Dolls Lay out the dolls for the parts in such a way that the dolls duplicate the flow of the schematic diagram. Do not be too

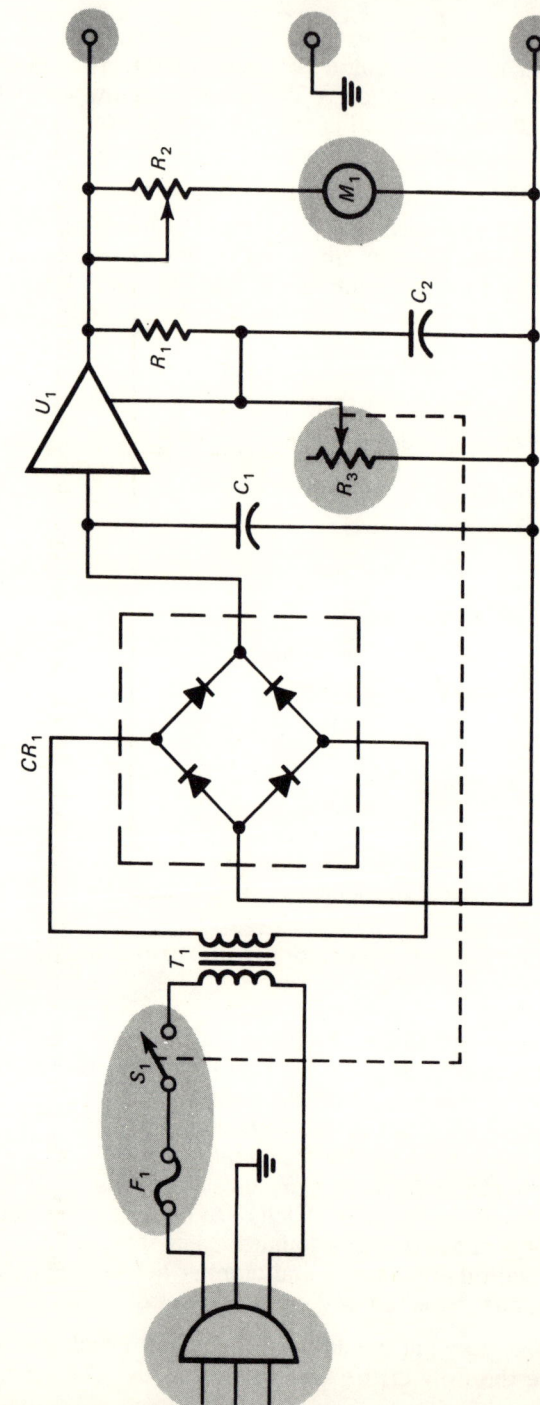

Fig. 3-12 Parts to be mounted off the board are shaded on the schematic diagram.

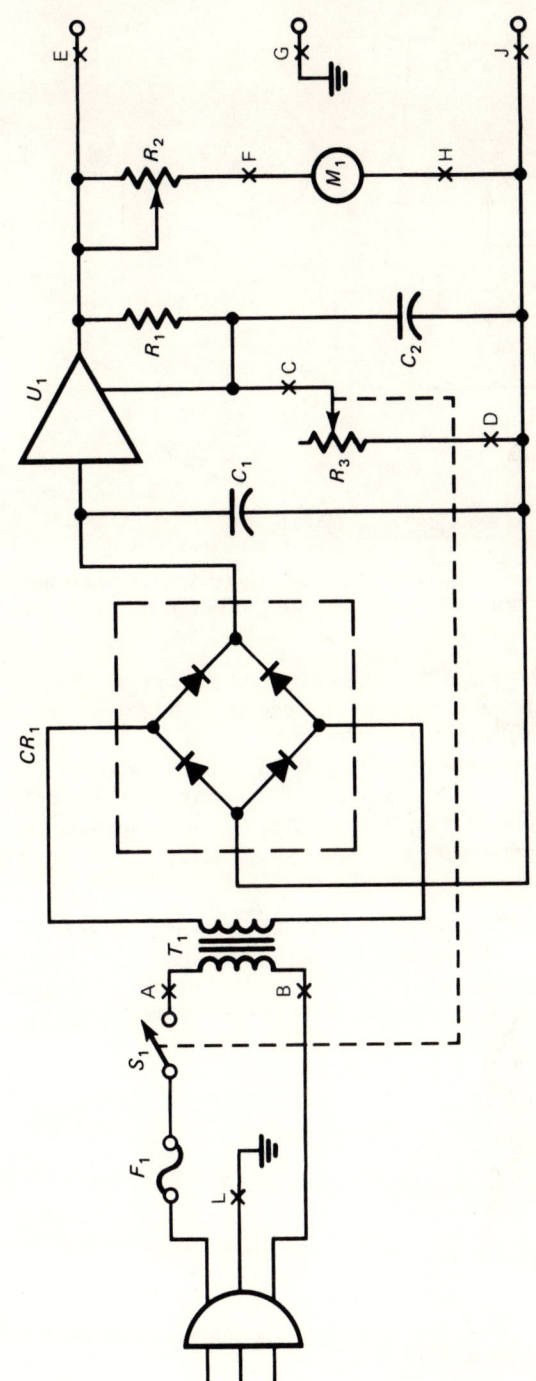

Fig. 3-13 This schematic shows off-board parts. An x is used to indicate wiring connected to off-board parts.

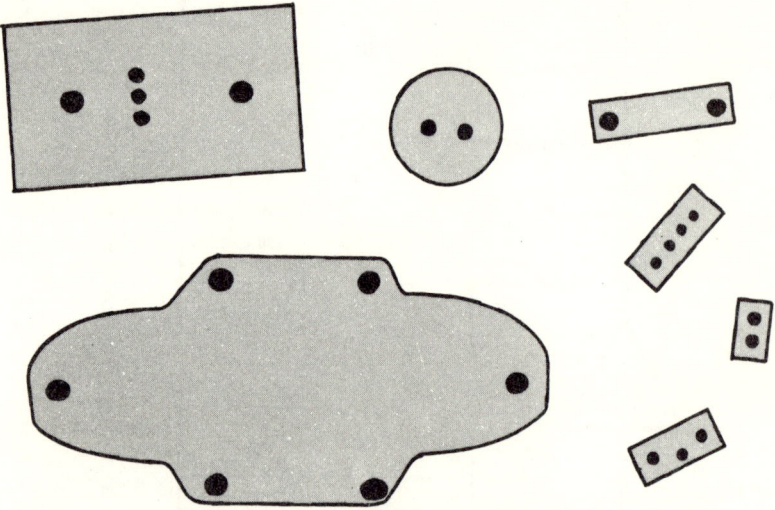

Fig. 3-14 A set of component dolls used to lay out the circuit board for the variable regulated power supply described in this book.

concerned about exact part and lead placement at this time. This procedure will probably make a relatively large board layout. If you are satisfied with the placement of the parts, and if you have room enough to fit everything into the cabinet, you may wish to plan the lines for the electric connections next. However, you may want to try to fit the parts into a smaller board. This will give a more compact package and reduce the cost of the project since you can buy a smaller board and cabinet.

Saving Space Sometimes during the layout of the circuit board space is at a premium. One way of saving board space and thus reducing the overall dimension of the board is to mount some components vertically (see Fig. 3-15). Diodes and resistors can easily be mounted in

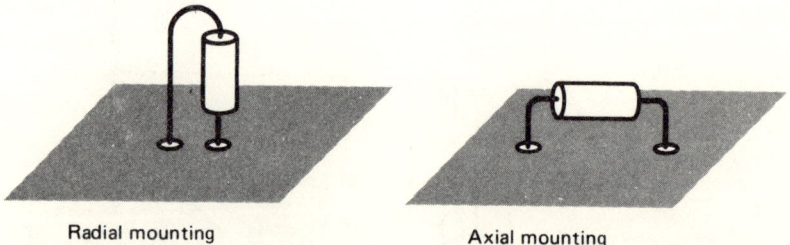

Radial mounting   Axial mounting

Fig. 3-15 Small parts, such as resistors or diodes can be mounted radially to take up less space on the board.

a "hairpin" shape on the board; just bend the leads as shown. Capacitors are manufactured with leads at each end (*axial lead arrangement*) or with the leads at one end of the unit (*radial lead arrangement*). Choose the type that best fits the overall design of the board you are developing.

Double-Sided Boards
Another way of reducing space is by using a double-sided circuit board. The procedure can be rather complicated because of the need for careful alignment of top and bottom artwork. If you decide to use a double-sided board, you will need to envision the artwork for both sides of the board. A double-sided board is shown in Fig. 3-16. Most boards of this type are laid out for use of photosensitive processes, discussed in Chap. 5.

Artwork If the decision is made to use a double-sided board, two pieces of artwork are required. They can be laid out the same as a single-sided board. The artwork for the underside of the board is done in reverse of the top side. Use of an overlay system is recommended for this process. In fact, it is easiest to use a backlighted worktable, such as a photographer's light table. Be sure to identify the edges of the artwork so that you will be able to align them again later on.

A heavy line representing the dimensions of the board is drawn on the tracing paper. It may not be the final board size, but it gives something to start from on your final layout.

Make up a working area for the board layout. Take a piece of graph paper. The size of the boxes is not critical, but try to use paper with 0.10-in spacing for the graph. Place a piece of tracing paper over the grid paper. You may wish to use a piece of cardboard as a solid backing for these papers. Make a hinge out of masking or mending tape for the tracing paper. This allows you to lift up the tracing paper to make adjustment without getting the tracing paper out of alignment.

Using Dolls The dolls you have made can be placed on the tracing paper or directly on the grid paper (Fig. 3-17). The purpose of the grid paper is to help lay out parts and connecting wires. Place the precut dolls on this package in position on the grid or tracing-paper work-

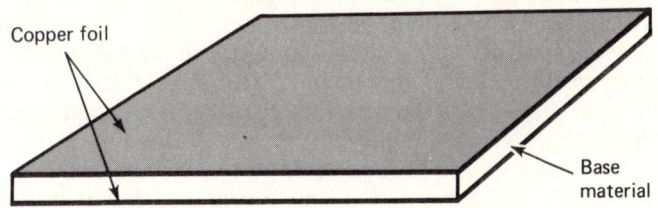

Fig. 3-16 Double-sided boards have foil on both sides of the base material.

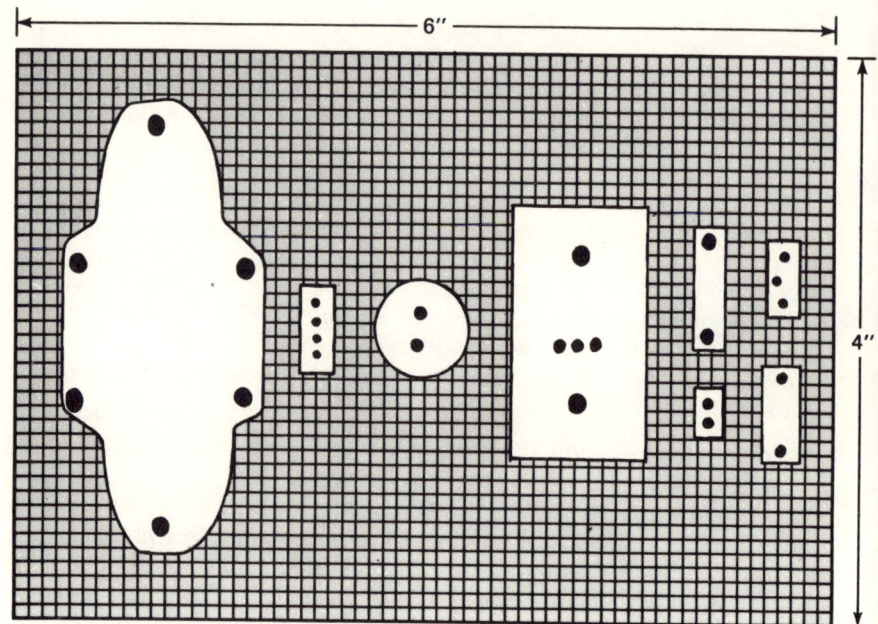

Fig. 3-17 Dolls are placed on the grid paper when laying out the board.

sheet. Next, use your judgment to decide whether the size of the board is acceptable or you want to try to reduce the overall dimensions. In the situation illustrated it was decided that the board dimensions could be reduced. A rearrangement of the parts is shown in Fig. 3-18. The overall objective of separation of input and output is next. Some of the parts are rearranged in order to reduce the size and cost of the project by moving some of the parts to new positions on the board. The electrical integrity of the circuit is maintained in this modified layout.

Conductive Paths The step after positioning each component to your satisfaction is to connect the parts to the power supply. Lines are used to indicate conductive paths. Each line on the schematic is checked against the conducting path on the board layout. Occasionally it is necessary to reposition one of the components in order to follow the guidelines given earlier in this chapter. (It's a good idea to reread the guidelines at this point to be sure they have been satisfied.) Remember that they are only offered to give you some direction in making the layout. You may find it is impossible to meet every single guideline. Use your own judgment. If the end result is a circuit board with a circuit that functions as it should, you have been successful.

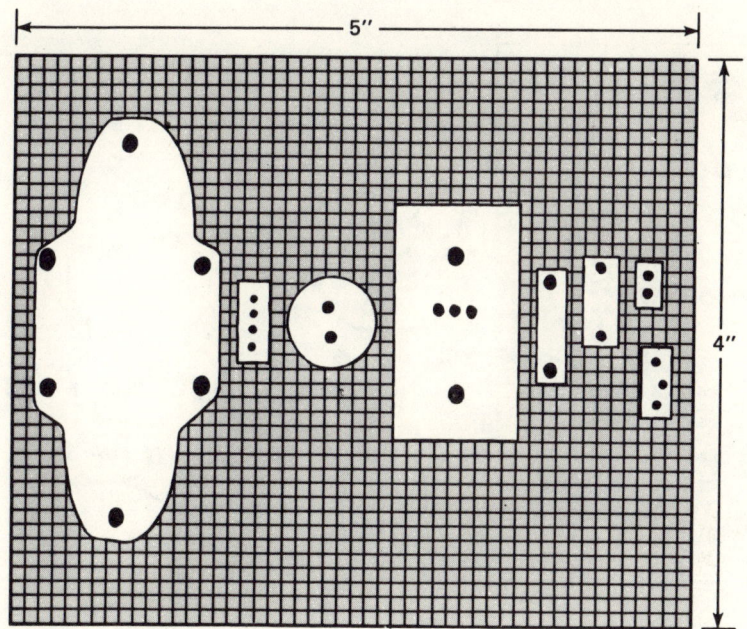

Fig. 3-18 The same dolls rearranged to make a smaller board layout.

Etched Areas Still another consideration is the area to be etched on the board. Some board designers try to remove all copper areas which are not used as conductive paths. Other designers leave copper in areas which are not used, as shown in Fig. 3-19. Using the second method cuts down on the etching time and makes the etching solution last longer because less copper is mixed with it. The choice is left to the designer.

EXAMPLE LAYOUT

A final layout for the power-supply project is shown in Fig. 3-20. This board uses component parts available from both local and mail sources. A layout of the parts placement and the conductive paths is given in the illustration. This layout follows the guidelines as much as possible for a board of this size. Extra copies of this layout are given in Appendix B to be used in later board developmental work.

Now is an excellent time to double-check your work. Consider each of the following:

1. Does the layout agree with the schematic diagram?

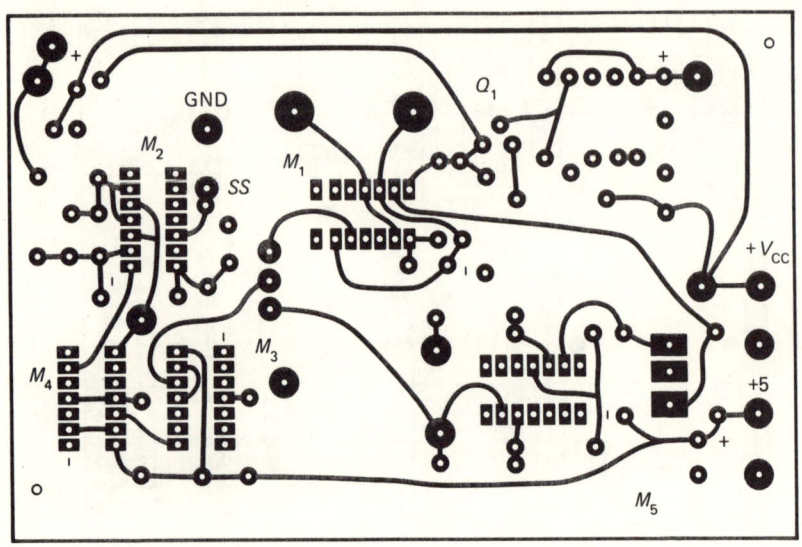

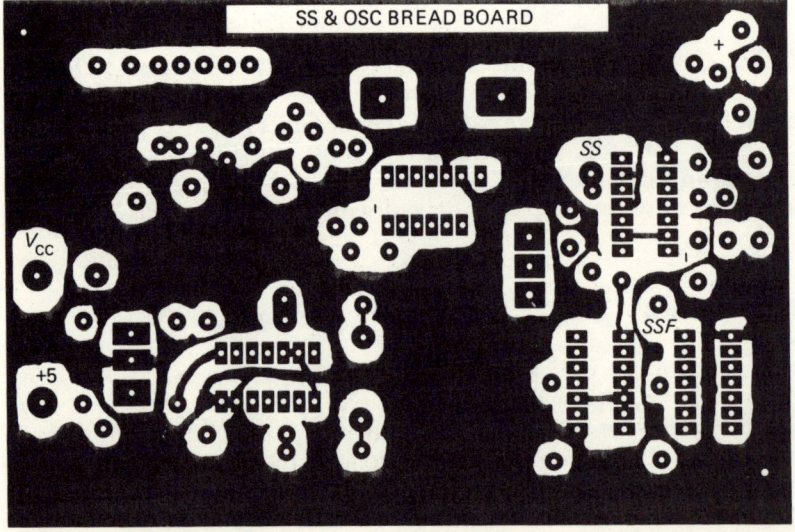

Fig. 3-19 Examples of standard board design and minimum-etching board design.

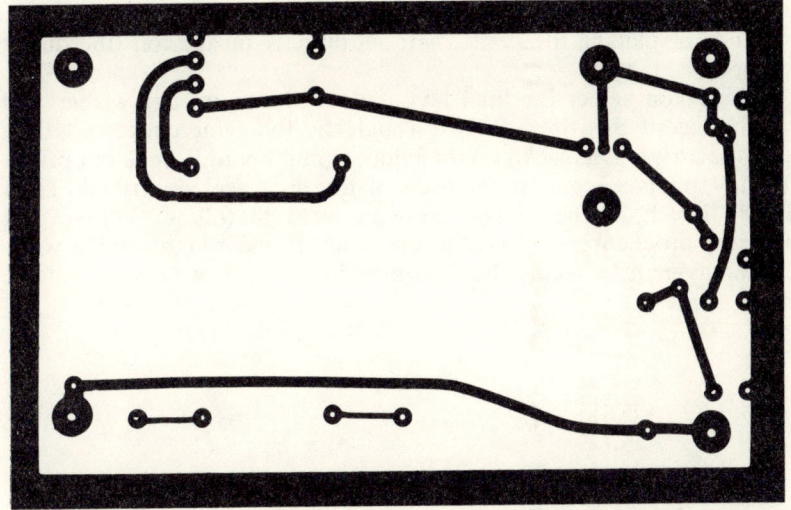

Fig. 3-20 Final board layout for the variable regulated dc power supply.

2. Do the physical dimensions agree with the cabinet and board size?
3. Are all parts found on the schematic on the board?
4. Have connections been identified for parts mounted off the board?
5. Are all parts of the proper size and scale (1×, 2×, or 4×)?
6. Are allowances made for such things as heat sinks, hardware, and mounting devices?
7. Can adjustments be made to controls and the like after parts are mounted (access)?
8. Is a separate mounting pad available for each component lead or terminal?
9. Do all conductors follow good design practice?
10. Are jumpers or crossover wires kept to a minimum?
11. Is there enough space between components and conductors to eliminate short circuits?

If you can answer "yes" to the above questions, you are well on your way to having a good circuit board.

The result of your efforts is a working drawing of the circuit-board layout. The final artwork will vary depending upon the method, equipment, and material selected for use. Two basic approaches are used: developing the circuit-board pattern using photographic methods or placing the resist material directly on the foil (the direct method).

Which you select for final layout depends upon your wishes and capabilities at this time. Read through the following chapters telling how the artwork is placed on the blank circuit board. Check out prices for the supplies required for each of the processes described. Then decide which of the techniques you want to follow. Try several different procedures to find the one you are most comfortable with. The enjoyment of seeing the final board develop is well worth all the effort. Read on!

4

Direct Layout Procedures

After the artwork has been completed, several more steps remain. The artwork has to be transferred from the paper layout to the surface of the blank board, which must be clean and able to accept the artwork. Next, the board is placed in an etching solution. It chemically removes those portions of the blank which are not protected. The result of these steps is an electronic circuit fastened to a supporting base material. Figure 4-1 shows the basic board without any mounted components. Holes for component parts are drilled next, and the parts are mounted and soldered in place. This completes the production of the board.

The board can be prepared for the etching solution in several ways. Some use light-sensitive materials to form *resist,* which is not acted upon chemically by the etching solution. It therefore protects the surface beneath it from the action of the etchant. Other methods use a resist that is applied directly to the surface of the board. These two methods are called the *photosensitive process* and the *direct process.* This chapter discusses the direct layout process; photographic techniques are discussed in Chap. 5.

BOARD PREPARATION

The surface of the blank copper board must be prepared to accept the artwork. This surface has to be completely free of oils, grease, and any other foreign substance for etching to be successful.

One of the easiest methods of cleaning the surface of the copper is to scour it with a mild abrasive material. There are special cleansing agents available, but common household scouring powder works very well. Proceed as follows. Wet the copper surface under running water. Sprinkle a small amount of scouring powder on the copper surface. Using a damp cloth or paper towel, rub the surface of the copper until it

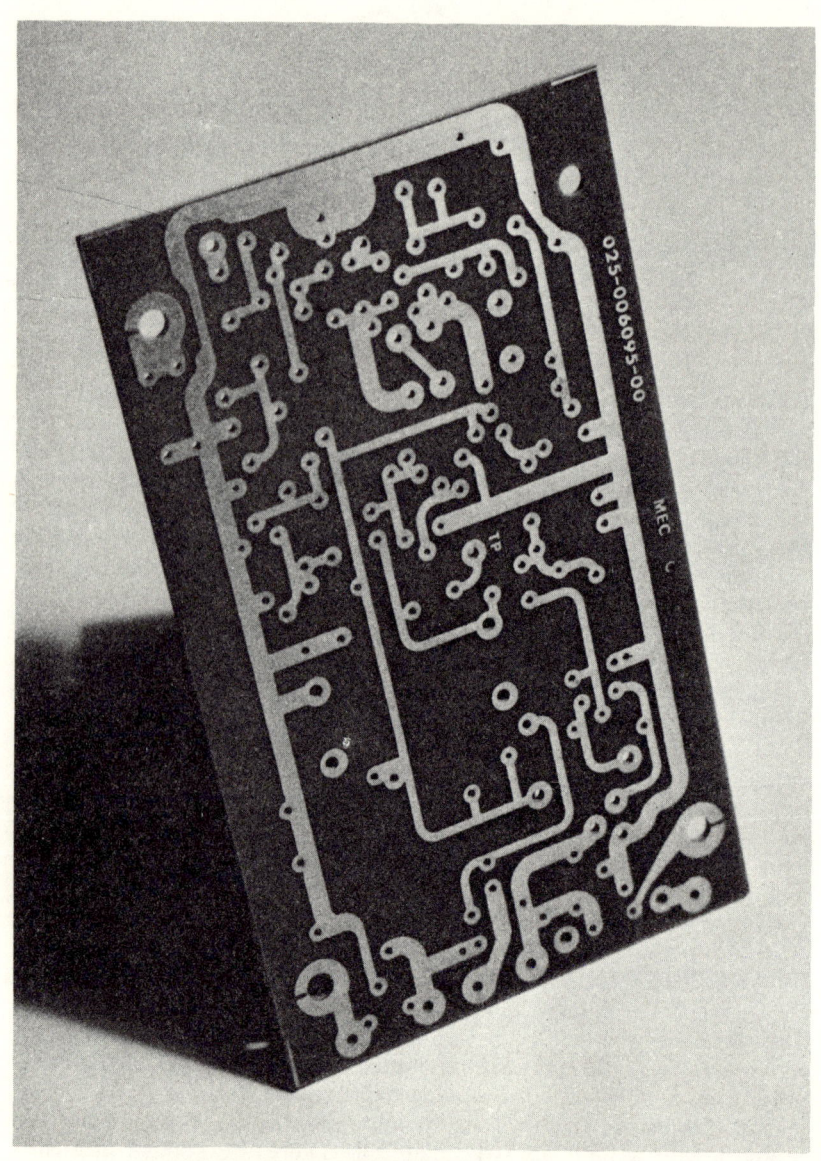

Fig. 4-1 A completed circuit board. Holes for parts are drilled. The board is ready for the parts to be mounted.

is shiny bright. Don't rub so hard that the copper is removed from the base material. Use a light pressure. Be sure to scour every portion of the copper surface. Most foreign materials responsible for a bad board are on the surface of the copper. They are easy to remove with this cleansing process.

After scouring the board, rinse it under running water to remove the abrasive material. *Be careful not to touch the copper surface with your fingers.* The oils on your body will contaminate the board surface. These oils are some of the foreign materials you have just removed, and you don't want to put them right back on the copper again.

Hold the board by its edges when handling it from now on until it has been etched. You can also hold it on the base side without touching the copper. Water on the surface of the copper is removed by standing the board on one edge and letting the water drain off or wiping the copper with a lint-free cloth. If any lint accidently remains on the copper surface, remove it before going on to the next step.

DIRECT RESIST MATERIALS

Methods of laying out a resist directly onto the copper surface of the board include using commercially prepared tape and dots, nail polish, paint, rub-on forms, and ink. Some of these methods are easier than others. Some are less expensive. Some make better-looking boards. Read through the descriptions and select the one that best fits your talents and finances. Try several different methods for a fair comparison.

Tape-and-Dot Process

One commercially available set of materials uses precut and shaped tapes and dots (Fig. 4-2) made of paper or plastic. These materials resist the etchants. Catalogs (see the sources listed in Appendix C) present a wide variety of sizes and shapes (not only dots) of these art forms. Tape-and-dot kits are available in $4\times$, $2\times$, and $1\times$ sizes. The larger sizes are used for work that will be reduced to exact size photographically (see Chap. 5). Only the $1\times$, (same size) are used in the processes discussed in this chapter.

How do you get the layout on the copper to match your artwork? Use your good judgment of spaces and distances and always refer to the artwork you have prepared. The companies that make the tape-and-dot sets also sell templates, or stencils, of the same shapes and sizes. When you have a template matching the tapes, dots, and other shapes you plan to use, drawing the artwork becomes much simpler. Templates were shown in Fig. 3-7. Spacing of components and lines is fairly critical but need not be an exact science. Use the artwork as a guide and place the dots where they appear on the artwork. Remember

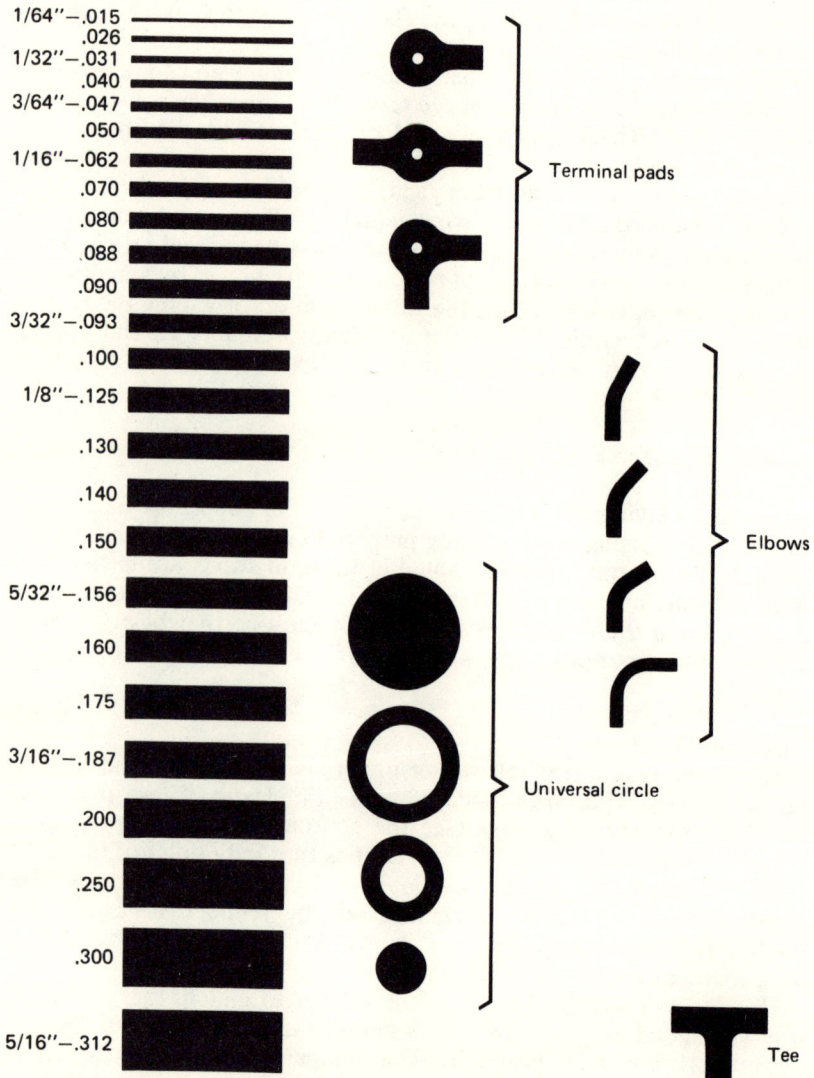

Fig. 4-2 Commercial tape-and-dot shapes available for use in board layout.

that they represent the pads used as mounting pads for component parts. The position of lines used as conductive patterns can be slightly different from that on the artwork.

Another way of solving this problem is to use carbon paper to transfer the artwork to the surface of the board. Place a sheet of carbon paper (the kind used in offices for making copies of letters) directly over the cleaned copper foil. Place the artwork on top of the carbon paper. If necessary, trim both the carbon paper and the artwork to the size of the board. Masking tape will hold the package together. Use a ball-point pen or a hard pencil to trace the artwork. Don't use any more pressure than you need—just enough to make the carbon paper work. It is only necessary to rough in the artwork. When you remove the artwork and the carbon paper, you should see an outline of the artwork on the copper. Do not be concerned if some of the carbon did not transfer. There should be enough information on the copper to guide you in placing the tape-and-dot forms.

The procedure for removing the dots and tape from the backing paper and transferring them to the artwork or copper foil is shown in Fig. 4-3. An artist's knife is used to lift one piece gently from its backing material. Using the knife simplifies final positioning of the adhesive-backed piece. You can see better where the artwork is being placed when you use a knife blade. Fingers also stick to dot-and-tape forms, and the results are messy and frustrating.

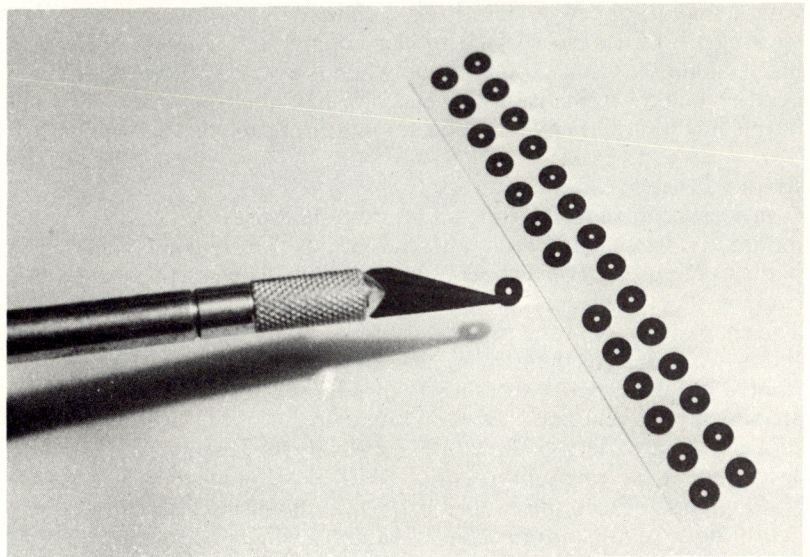

Fig. 4-3 Use an artist's knife to lift the artwork from its backing paper and place it on the board layout.

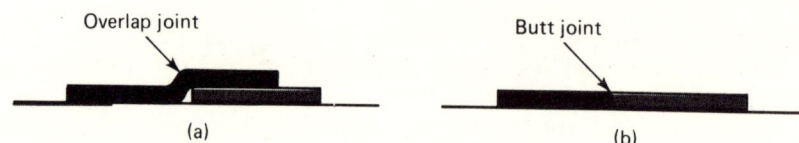

Fig. 4-4 (*a*) Overlap joints tend to leave areas where etchant can cut through the tape. (*b*) A butt joint is better.

Using the artwork as a reference, select a form of the proper shape. Stick it on the copper foil directly over the place indicated by the tracing carbon. Remember to use an artist's knife, not your fingers. After the forms are in place, connect them with the tapes. Use the same procedure as you used with the forms. Run a tape of the correct width between the appropriate forms. Be sure you have a good clean connection wherever two pieces of the resist material join. Proper and the improper ways of joining two pieces of artwork are illustrated in Fig. 4-4. When an overlap occurs (Fig. 4-4*a*), the etchant can work its way under the joint and eat away copper where it shouldn't. The result is a board whose conducting paths have a few extra breaks. A better method is to cut the tape with a sharp knife at the junction before sticking it on. The two edges should meet without a gap. This is called a *butt joint* (Fig. 4-4*b*).

After the forms and tape have been positioned they must be burnished. This simply means rubbing the resist material on the board with a smooth-tipped device. It ensures that all the dots and tapes are sticking firmly to the surface of the copper foil. Failure to burnish means that areas that should be foil when the board is completed have been etched away instead. The cap of a ball-point pen makes a good burnishing tool, but any kind of a smooth firm rubbing tool will work.

Rub-on Process
A process similar to the tape-and-dot method uses dry-transfer resist patterns, also used by commercial artists for making signs. The patterns are printed on a sheet of clear plastic. The plastic is stuck to a backing sheet to protect the patterns from being removed accidentally. The shape or form of the patterns is limited only by the imagination of those who design them. Samples of the shapes used in making printed-circuit boards are shown in Fig. 4-5. They include actual shapes of pads and lines for interconnection.

Using these patterns is very similar to using the tape-and-dot process, described in the previous section. Start by removing the protective backing sheet. Place the proper pattern in the correct position on the board. *Be sure the board is clean and dry.* Using a ball-point pen, rub over the pattern with a firm pressure. Any hard, round device can be used as a rubbing or burnishing tool.

When you have finished rubbing, peel the sheet away from the board

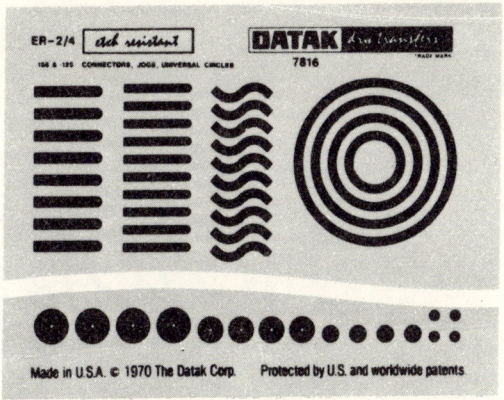

Fig. 4-5 Rub-on art designs for board layout. (*DATAK Corporation.*)

and the pattern is left on the board. Put the backing sheet over the pattern and burnish again. Use as much pressure as you can apply with your hand to be sure the artwork is firmly in place. The end result is shown in Fig. 4-6.

Connect the dots and pads with the line patterns. With a little practice you can use only part of a pattern. In other words, you can rub a part of a circle or other pattern onto the surface of the board. This gives an almost unlimited variety of shapes for use in board layout work.

After the circuit has been laid out, burnish it again. Check to be sure that there are no wrinkles or breaks in the artwork. They will mean etching where you don't want it.

One word of caution. The patterns are very thin. Ordinary etching solutions have time enough to eat through them. The pattern manufacturer recommends using a high-speed etching solution. The various etching solutions are described in Chap. 7. High speed etchants are available from the manufacturer of the patterns and from other sources (see Appendix C).

Pen-and-Ink Process
Another method of putting a resist on the blank circuit board is to use artists' pen and special ink. Regular or India ink does not adhere to the surface of the copper. The special ink available for this purpose is called *resist ink*.

The artwork is prepared as before. Anyone using this method must have a very steady hand. The artwork is drawn with pen and special ink on the surface of the blank copper board. Start with the circles and pads to which parts are to be connected. When they have all been laid out, connect the dots and pads with lines for conductive paths. Do not

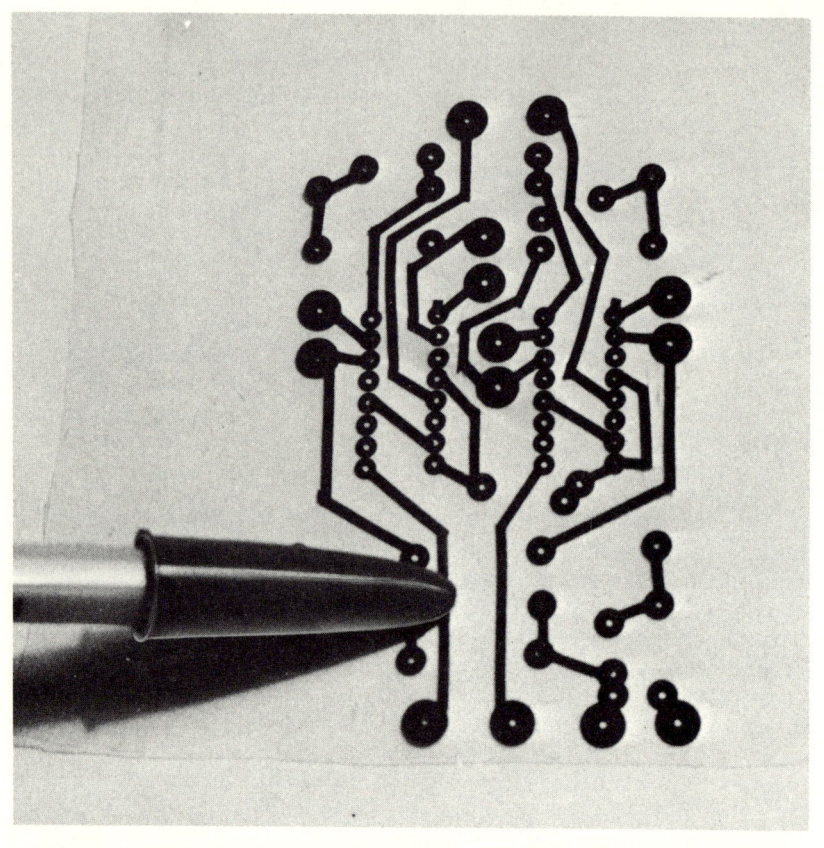

Fig. 4-6 Rub the artwork in place firmly after it is placed on the layout paper.

move the board until you are certain the ink is dry. After the ink is dry, check the board for breaks in the resist and touch them up.

The major disadvantages of this process are the time it takes and the skill it requires. Some of the more complicated patterns used today are difficult to lay out (this is particularly true for integrated-circuit mounting pads). They require a lot of time and a very steady hand. Try this method to see how well you can do it.

Rubber-Stamp Process
A variation of the pen-and-ink process uses commercial rubber stamps for the pad shapes. Rubber stamps are available in most of the common shapes used to mount transistors and integrated circuits. Some of these forms are shown in Fig. 4-7 along with the impression of the rubber stamp on the foil.

A special resist ink is required for the stamps. (It is the same resist ink used for the pen-and-ink process.) The rubber stamp is inked on a stamp pad using the resist ink. The stamp is placed in the appropriate place on the copper foil. Proper pressure is essential for this work. Too little pressure will not transfer the ink from the rubber to the copper. Too much pressure will cause only the outline of the stamp to be printed. A little practice will enable you to find the proper pressure to use.

Once the shapes are stamped on the copper, the connecting lines are drawn in. Either an artist's pen with resist ink or a special ball-point pen containing the resist ink can be used. (Removing resist ink if you make a mistake is discussed below.)

This method seems to be a good compromise between the pen-and-ink and the tape-and-dot methods. The cost of the set of rubber stamps is fairly low compared with that of prepared patterns, but the shapes available on a rubber stamp are somewhat limited. Skilled hobbyists can easily modify existing rubber stamps or carve simple patterns from blank rubber-stamp materials to produce unusual or special forms. Using rubber stamps saves time, and faithful reproduction of patterns is well worth the cost of the materials.

Other Processes

The direct method of placing a resist on the surface of a blank circuit board is limited only by the need for the material to resist the etchant. Many materials found in the average home can be used as resists, for example, nail polish and oil- or lacquer-base paints. When choosing a resist, always test a sample board in the etchant. Make up a nonsense board or (if you feel brave) a small working board using the resist material you want to test. How does the material act when the board is placed in the etchant? If the circuit holds and only the clear copper is removed during the etching process, you have a good resist material. If the resist is etched away with the foil, the test has failed and that material should not be used as resist. Use only materials that are successful and easy to use in preparing the layout.

Fig. 4-7 Rubber stamps are available in more complex forms. (*Rainbow Industries.*)

A FINAL THOUGHT

All the methods described in this chapter can be used successfully to make a resist pattern directly on the copper foil. The one you choose is limited only by your budget and the steadiness of your hand in drawing the artwork on the board. The methods described can be used to supplement each other. Try a little paint or resist ink to correct a minor problem. Be sure you have the necessary materials on hand to remove the resist in case you need to change the artwork before etching. Resist can be removed by a variety of methods, both chemical and mechanical. The chemical removers act as solvents for the resist. It would be wise to know which chemical removes the resist and have it on hand before you etch the board. After the board is etched the resist remover should be used. The solder will not stick to the copper if this is not done. Be prepared to do a complete job. Try samples first before starting the big all-important project. The success of the project is limited only by your care in developing the original board.

5

Photographic Layout Procedures

One of the hardest parts of laying out a board by the direct process is transferring the artwork onto the surface of the board. This is time-consuming and sometimes frustrating. Isn't there a better way?

The better method costs more than direct methods, but it is a lot simpler. It also produces equal or better results by forming a resist on the surface of the blank board photographically. The original artwork is used to make a mask. The mask is placed on top of a board prepared with a photoresist material. When a light is held over the mask and board, the resist material reacts to the light. The light is removed, and the board is put in a developing solution. This hardens the resist where it is exposed to the light. Other areas are not hardened and will etch away rapidly.

The following steps are involved in this process:

1. Lay out artwork

2. Make photographic mask (several methods discussed)

3. Prepare photoresist (or purchase prepared board)

4. Expose board

5. Develop board

6. Dry board

Each step is described in this chapter.

Equipment requirements are very simple. There is no need for photographic enlargers or a printer's copy camera. Most of the work is done under direct sunlight or a photoflood lamp. The only other equipment required is a glass frame to hold everything together during exposure and a glass tray in which to develop the board after it is exposed.

ARTWORK

Artwork requirements for the photographic processes are similar to those for the direct process. The same layout rules apply, the only major difference being the material on which the artwork is laid out.

Before discussing these materials, a clarification of some often confusing terms is in order. The terms *negative* and *positive* refer to the type of artwork and have nothing to do with the polarity of the operating voltage in a set.

The Negative
A photographic negative is illustrated in Fig. 5-1. It is the opposite of the artwork, which is also shown. Black surfaces appear in areas which would normally be white on the artwork. White, or clear, areas appear in places which are black on the artwork. The negative is a true opposite of the original artwork.

The Positive
A photographic positive is illustrated in Fig. 5-2. The positive looks exactly like the original artwork. It is clear where the artwork is white, or clear. It is dark in areas that are black on the original.

It is important to understand these terms. Some photographic processes require a negative mask, and others require a positive mask. This is the only difference in the process. Using the incorrect mask makes the final board surfaces just the opposite of what you intended.

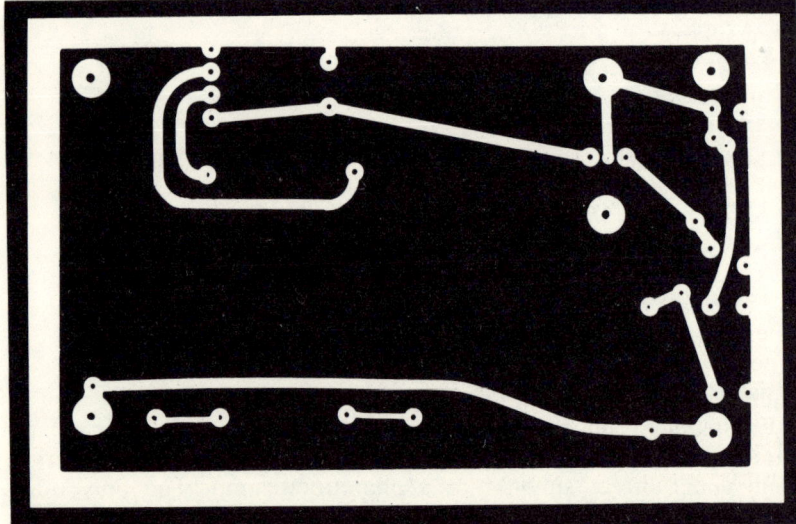

Fig. 5-1 The negative is the opposite of the artwork. Black appears white in this form.

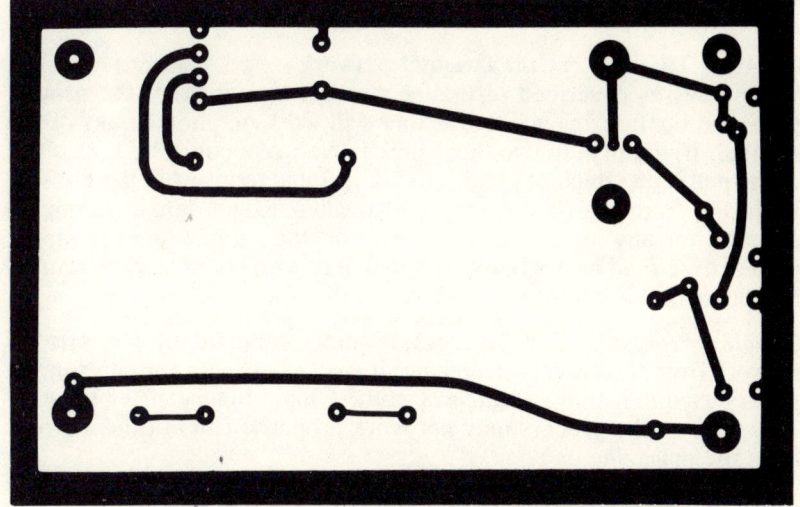

Fig. 5-2 The positive is the exact form of the artwork as it will appear on the circuit board.

Artwork Size

In laying out boards for commercial use the designer will use artwork that is 4× or 2× actual size. The completed artwork is reduced to 100 percent of final size by use of a printer's copy camera. If you have access to one of these cameras, by all means use the larger size layout. Since most of us do not have copy cameras, we prepare the artwork at 100 percent, or 1×, the final size. Either way, follow the steps described in this chapter.

If you use the larger size layout and plan to reduce the artwork photographically, *be sure that all your work is done on the same scale*. Include dots and pads of the proper size. Spacing between conductors must be enlarged as well as the size of the conductor. Component sizes will be either 4× or 2× exact size for layout purposes.

MAKING THE MASK

The photographic mask is a positive or a negative, depending on the type of resist material used on the board. Both positive and negative resist materials are available. The process used to make the mask is the same in either case. Most commercially available materials use a negative mask. The exception, the Excel Circuits process, uses a positive mask and is described below.

Mask-making processes fall into two categories: (1) the original artwork is destroyed as a part of the process, and (2) the artwork is not destroyed.

Processes That Destroy the Original Artwork
The processes described in this section lift the ink from the printed page of a book or magazine. Neither will work on photocopies of the original. If you plan to use these processes, make copies of both sides of the pages on which original artwork is found to insert in the book or magazine from which the artwork is taken. If you are planning to practice on any of these processes for the variable-power-supply project described in this book, you will find extra board layouts printed in Appendixes A and B.

Circolex Process* This process requires removal of the printed artwork from the book or magazine. It works best with smooth surface papers used in leading magazines in the United States. If the paper is very porous, the process may not work properly. Cut out the artwork from the magazine or book.

A flat, level work surface is required. Check it with a surface level or bubble level. Next, tape a sheet of paper onto the work surface. Tape it at the corners and wherever necessary to keep it perfectly flat. Be sure the paper is larger than the artwork you are working with. Fasten the artwork face up onto the paper in the same way.

The kit of materials provided for this process includes a set of sticks for making a frame. The frame is constructed on the artwork, leaving an inside border of about ½ in between the artwork and the sticks. Cut the sticks to length with a sharp tool. Lay them flat on the artwork. Hold the sticks in place with masking tape.

Chemicals are used in this process to make a film. It may shrink when it is cut loose from the frame. While this is not a major problem, even a reduction in size of 1 to 1½ percent will affect the spacing of components. If this is critical, lay a piece of screen wire about ½ in wide around the artwork before continuing with the process. The screen wire will keep the negative from shrinking when it is removed from the frame.

The chemicals provided with the kit require measuring and mixing. Instructions for both are included in the kit. After stirring for 1 minute pour the mixture into the artwork. Start at the center of the artwork and work toward the edges of the frame. If necessary the mixture can be "pulled" toward the edges with the stirring stick. Be very careful not to touch the artwork with the stick. Touching can damage it.

The curing, or drying, time for this process is 1½ hours. *Do not attempt to remove sooner even if it seems dry.* After curing, the plastic formed from the chemical mixture is cut away from the frame with a sharp instrument. The paper is still attached to the plastic. Holding the plastic under warm running water, peel off the paper. Remove any

*Based on information from Circolex Scientific Co.

whitish paper residue left on the surface of the plastic. Use only a light rubbing action. Hard rubbing may destroy the whole works.

After air drying, the negative is ready to be used. Take care not to scratch the surface of the negative. Store it in a cool, dry place. If necessary the negative can be retouched with a black marking pen. Check it to be sure that all connections are made and there are no bubbles or areas where the circuit is not complete.

LifTact Process Another lift-off technique is the LifTact process. It uses a plastic film to remove the printed artwork. Much less time is required for the total process than for the Circolex process (only about 15 minutes).

The transfer film is supplied with a backing sheet. The first step is to remove the backing. Cut the film to the approximate size of your artwork. Tape the artwork down on a smooth, flat, solid surface. Remove the backing material from the film and discard it.

The film is applied directly to the surface of the artwork. *It cannot be lifted or repositioned once it is in place.* Do it carefully the first time. The film is now burnished to make it stick. This process also removes air bubbles. Air bubbles do not adhere to the ink and cause gaps in the artwork if they are not removed.

Next, put the artwork in a pan of warm water for about 15 to 20 minutes. When the time is up, very gently rub or peel the paper off the film. Use the tip of your finger. The paper will start to ball up and wash off very quickly. Once the paper is removed, wash the film under warm running water. Allow it to dry in open air. The film is now ready to be used.

PCP Transfer Film Process The PCP Transfer Film process is similar to the LifTact process, and procedures are the same. Both processes are simple and easy to use. The materials required to produce the mask are few and can be obtained from the sources listed in Appendix C.

Processes That Do Not Destroy the Original Artwork

Excel Circuits Process A process that is rarely described in the literature is one developed by Excel Circuits. It uses a board with a positive photoresist on it. Using a positive photoresist has several advantages for the hobbyist, for example, making the artwork easy.

The artwork is made by placing a sheet of Mylar (plastic) over the top of the artwork model. Artwork from a book or a magazine can also be used. Commercial tape and dots (see Chap. 4) are placed on the Mylar sheet. Ordinary black ink can be used to draw the artwork, correct errors, or connect pads. The completed Mylar sheet becomes the artwork mask.

Another advantage of this process is that the image on the exposed board is very easy to see and inspect. Areas which will be etched become green. Areas to remain on the board are blue. Traces of blue in the background of an exposed board indicate incomplete photographic exposure. When this occurs, the artwork and board are put together in their original position and reexposed.

A third advantage is that all chemicals are water-soluble. There are no toxic fumes. This reduces the need for ventilation required when working with other process chemicals and boards. It is not necessary to bake these boards after developing in order to set the resist, as required with other processes.

The instructions for use of this process are very easy to follow. A positive of the artwork is placed directly on the surface of a pre-sensitized board, available only from Excel Circuits Co. It is not possible to buy the resist and coat your own boards. If you wish, and if the equipment is available, you can make a transparency of the original magazine artwork and use it as a positive of the artwork.

The photoresist-coated board should be worked in reduced room light. Fluorescent lights will spoil the boards by exposing them in about 10 minutes. Avoid them by using standard light bulbs wherever possible. Place a sheet of glass on top of the artwork. This forces the artwork and the board into close contact with each other.

Exposure is made with a no. 2 photoflood lamp. Time for exposure is 10 to 15 minutes. To speed up the process you can use a 275-W ultraviolet sunlamp instead of the photoflood lamp. Exposure time with the sunlamp is 3 to 5 minutes. If the board does not appear to be completely exposed at the end of the prescribed time, expose it longer. This will not hurt the resist or the board.

A kit of materials available from Excel Circuits Co. contains boards, artwork materials, and both developing and etching chemicals. Individual replacement supplies can be purchased from the company when necessary. The chemicals are shipped in concentrated form and must be mixed with water for use. The developer is mixed with water at a temperature of 70 to 80°F. Hot water will remove the resist from the board, so be careful. Extra developer can be stored in a capped bottle after mixing.

Pour some developer into a glass or plastic tray. Place the board in the tray face up. Gently rock the tray for about 2 to 3 minutes. When all the resist in the background has been removed, the board is done. Take it out of the developing solution and rinse it under warm water. Let it dry. When dry, the board is ready for a final inspection and any minor touch-up required to repair errors or other goofs. The board is now ready for etching. The process described in the Excel kit must be used.

3M Color-Key Process*   The artwork for this process may be made of

*Based on information from 3M Industrial Graphic Division, 3M Corp.

any material *as long as the material does not allow light to pass through it*. Artwork may consist of tape and dots. It may also be a drawing done with pen and ordinary black ink.

The key to this process is a film produced by the 3M company primarily for use in the printing industry. It is used to make the positives and negatives required for offset printing. The printed-circuit industry can take advantage of these techniques because making an offset printing plate is very similar to making an electronic printed circuit.

First decide what type of mask you want to make. This Color Key process can produce negative transparent, positive transparent, and negative opaque film masks. The type of film required for the printed-circuit process is *negative opaque*. When ordering this film, be sure to specify that you require the negative opaque material.

You will also need a source of ultraviolet light, a frame in which to hold your work, and a developing surface on which to work. The light source can be a 500-W photoflood lamp. Set it up so that it is about 18 in away from the film copy area. If you live in an area with a lot of direct sunlight, you can use it instead of the photoflood lamp. A unit capable of being used for exposure of the film and later the photosensitized circuit board is illustrated in Fig. 5-3. The unit has a wooden base and an upright support made of pipe. The photoflood lamp is

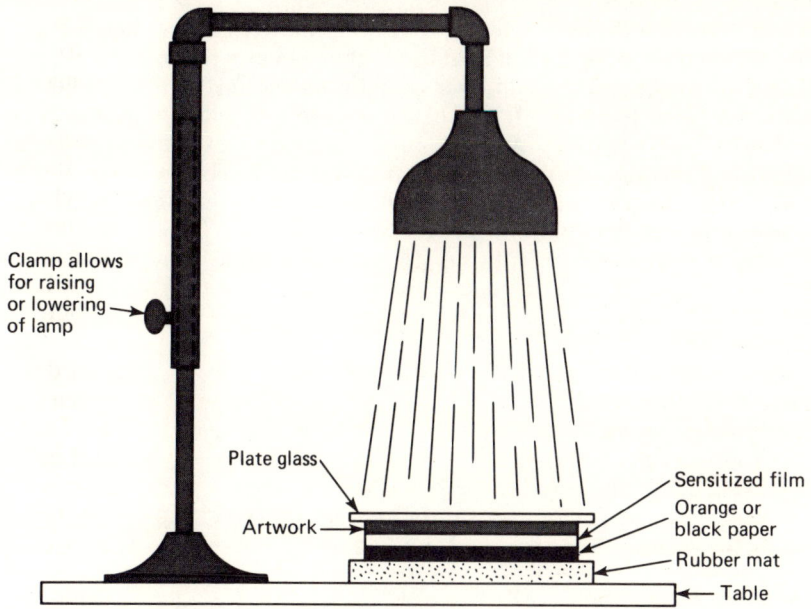

Fig. 5-3 A homemade light unit used to expose the photographic mask.

clamped onto the pipe at the appropriate height. A phototimer is necessary for accurate timing. This versatile unit will be handy for most photographic exposure requirements.

A word of caution about photosensitive, or light-sensitive, materials. Light from the source reacts with the chemicals on the film. The film is not smart. It cannot tell the difference between good and bad light, that is, light you want to have hit the film in certain spots and light coming accidentally from unwanted sources that can ruin the film. Most manufacturers of photographic supplies tell you what kind of light (if any) you can work in *before* you are ready to expose the film. It ranges from red, as seen in darkrooms, to yellow or normal room light. Be sure to read the instructions carefully or you may ruin a whole batch of film.

3M Color-Key film can be used in normal room light for short periods of time. It can also be used under the red or yellow safety lights recommended for photographic processing. To be on the safe side work away from the direct sunlight and, if possible, turn off the overhead lights in the room. Light coming in a window during the day will provide a lot of work light and not damage any film.

Start the exposing process by making a sandwich consisting of the artwork, the Color-Key film, and a backing sheet of orange or black paper. The exact order of the sandwich layers is important. Be sure to read the instructions for the Color-Key process and know them exactly.

An exposure frame is used to hold the sandwich firmly together. If the film and artwork are not held tight, light can creep into areas where it doesn't belong. This spill-over of light makes lines on the negative narrower than intended. It can cause one area to bleed into another, giving a bad negative. A sketch for a wooden exposure frame is illustrated in Fig. 5-4. You can use an inexpensive picture frame as the foundation for this unit. It can be hinged at one end and made an integral part of the light unit, as shown.

Exposure time for the film in a homemade frame runs from 2 to 4

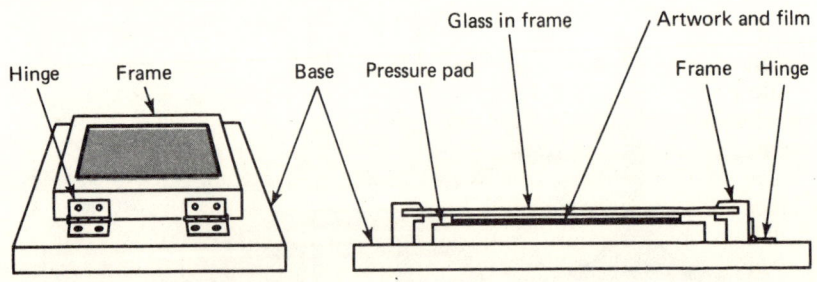

Fig. 5-4 The exposure frame is used to hold the positive flat against the film.

minutes. Time will vary due to differences in the type of supporting material used with the artwork. The best thing is to try various times in order to see which ones work best with your materials.

After the film is exposed it must be developed. Each type of photosensitive material has its own kind of developer. The 3M Color-Key film is no exception. Buy a bottle of developer when you buy the film, and follow the developing instructions supplied with the film. The film is placed on a sheet of glass in a tray. Developer is poured onto the sheet. A block of wood covered with a wiping pad is used to spread the developer around on the film. This removes most of the background coating on the film and the image begins to appear. Reverse the pad on the block to give a clean surface and complete the process. As a final step, rinse both sides of the sheet of film under room-temperature water. Blot the film dry with an absorbent paper. Your film is now completed. A negative of your original artwork should be on the film and ready for the next step.

Copy-Camera Process Another way of producing a negative is to use a printer's copy camera. If you do not have access to such a camera, you can make arrangements with either a printer or a printed-circuit-board manufacturer to have a negative made. Companies advertising in various trade and electronic journals will make a negative if you provide the artwork.

PHOTORESISTS

The methods described in this chapter will enable you to produce a negative for use as a mask. The next step is to buy or prepare a blank circuit board covered with a photosensitive resist. The photographic negative is placed on top of the prepared board. When exposed and developed, the photoresist forms a resistive positive on the board. The positive is in the shape and form of the conductive paths required for the finished board.

Board Preparation
The blank circuit board can be cut to its final size at this time, but some people prefer to leave it larger and trim it after etching. Rough edges are cleaned up with a file. The copper surface is cleaned with a commercial cleansing solution or kitchen scouring powder. Wet the surface of the copper, sprinkle some of the powder on the wetted surface, and use a damp cloth or paper towel to scrub the copper. A small handbrush may also be used. This cleaning is to remove all traces of grease, oil, or copper oxide from the board. Be careful to use a mild abrasive. Take care not to put any deep scratches in the surface of the copper.

Rinse the board under water. If the board is free from oils and oxides the water will appear to roll off the copper in sheets. If any contaminants remain, the water will form beads in these areas. After the board is cleaned, dry it with a lint-free cloth to remove all surface moisture. Then dry the board in a 120°F oven for about 10 minutes. Remove the board from the oven and let it cool to room temperature. Take care to handle it only by the edges or bottom so that you do not get any oil from your hands on the surface of the board. The photoresist material is applied next.

Resist Application
The photoresist is a light-sensitive liquid. Work in an area of subdued light or use yellow safety lights. The resist is available in liquid form or in a spray can. It is rolled, spun, or sprayed onto the surface of the copper. Most commercial processes use rolling or spinning to coat the board, but they require large facilities and are easily contaminated. The hobbyist is better off using spray cans.

Stand the clean board against a vertical surface as illustrated in Fig. 5-5. Shake the can thoroughly. Test the spray before coating the surface of the board. Hold the nozzle down for 1 or 2 seconds to clear it and allow for complete atomizing of the resist. Keep the nozzle about 8 or 10 in away from the board when spraying. Most manufacturers suggest starting the spraying at the bottom of the board. Spray horizontally and move upward at the end of each pass with the spray. This will provide an even coating on the board.

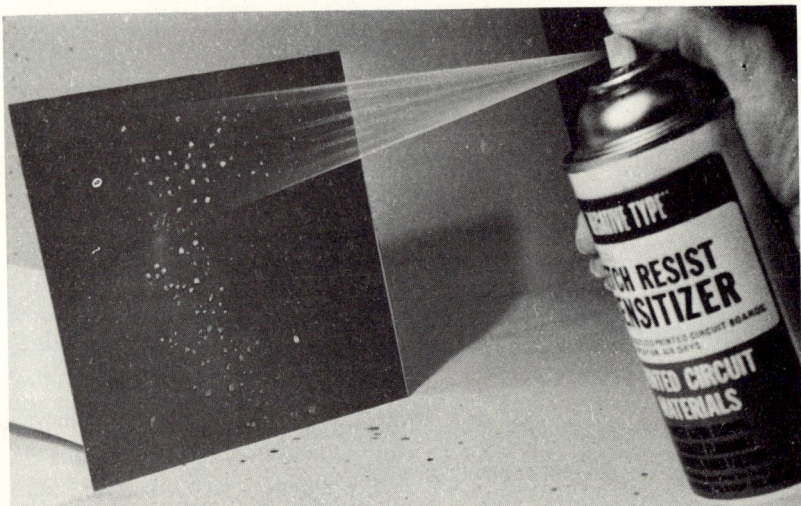

Fig. 5-5 The clean board is placed against a vertical surface in order to spray resist onto it.

After spraying lay the board in a horizontal position for about 1 minute. This allows the wet resist to level off uniformly over the copper surface. Then dry the board in an oven at 110 to 115°F for 10 to 15 minutes. This sets the resist and removes any solvents in the resist which keep it from sticking to the copper surface. As the resist hardens, it is very sensitive to light. Take care that it is not accidentally exposed to light. The sensitized board can be stored at room temperature in any lighttight container.

EXPOSING THE BOARD

The photosensitive resist is exposed just the way the negative was. Use the same exposure unit shown in Fig. 5-3. In this procedure the photoflood lamp is moved closer to the board. About 12 in is good for board exposure. Working in subdued light, place the board on the table with the sensitized side up. Center it under the lamp. Carefully place the negative of the artwork on the surface of the board. Be sure that the negative is placed so that it covers the surface of the board. Try to align it with the edges of the board. Place a piece of glass over the negative and board. The glass should be heat-treated so that it will not crack when exposed to the heat of the photoflood lamp. The weight of the glass will hold the negative tight against the surface of the board.

During the exposure period of 5 to 6 minutes the photoresist reacts to the light. This reaction sets the resist. If the negative is moved during exposure, the outline of the conductors on the board will also shift. If the negative is not tight against the board, light will leak into adjacent areas. Both result in a board of poor quality. Take care to avoid these situations.

DEVELOPING THE BOARD

Once the board is exposed, it is developed. The developing solution must be compatible with the resist. (Buy the developer and resist from the same source to be sure that they work together.) The developer is poured into a tray of glass or metal because most developers react to plastics. This reaction prevents the resist material from setting properly and makes a big mess if the developer starts to leak through the plastic container.

Place the board in the developing solution. Be careful not to touch the copper side of the board. Developing time is anywhere from 40 second to 1 minute. Gently rock or agitate the tray during development to keep the fluid moving over the surface of the board. Remove the board from the tray and let the developer drip back into the tray. Stand the board up on a piece of paper towel so that any developer on the surface of the board can drain onto the towel. You should be able to see a light outline of the circuit on the surface of the board. The developer can be poured back into its container if it has not become contaminated. Normal developing will not hurt the solution.

Once the board is developed, it is no longer sensitive to light. Normal room light can be used for all future work on the board. The resist is soft when the board is first removed from the developer. Make sure the pattern on the board is not damaged by careless handling.

For those who want to see how the board will look a spray dye is available. It penetrates the hardened resist on the board. Using the dye will show up the pattern on the board. This step is not required, but it may be reassuring to be able to see the final pattern at this time.

FINAL WORK AND DRYING

It may be necessary to make minor corrections to the board at this time. If you have worked carefully and planned ahead, no major corrections will have to be made. Careful use of an artist's knife will remove any excess resist material. A paint brush and a liquid resist are used to make any additions to the circuit. After all corrections have been made, the board is given its final preparation for the etching tank. The resist must be hardened by drying. This process can be done at room temperature. It is also done in an oven at 200 to 250°F for 30 minutes.

DOUBLE-SIDED BOARDS

The double-sided board has copper on both sides of the supporting base material. Each side has its own artwork. The component parts are mounted on the side with the least circuit conductor wiring on it. A double-sided board is shown in Fig. 5-6.

Laying out a double-sided board is similar to laying out a single-sided board. The simplest way to identify the two sides is to use two colors of pencil. Red and blue are good contrasting colors. Both colors can be laid out on a single sheet of tracing paper, or a separate sheet can be used for each side of the board. Try to keep most of the conductors on one side of the board. Otherwise the procedure is the same as before.

Artwork

A double-sided board requires separate pieces of artwork for each side of the board. If you have a camera and some lens filters, you can use the two colored pencils (or colored commercial artwork) and do all the layout on one sheet. The filters will permit either the red or the blue to be photographed. This produces masks for both sides of the board.

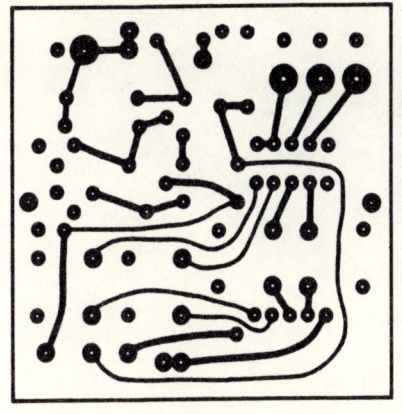 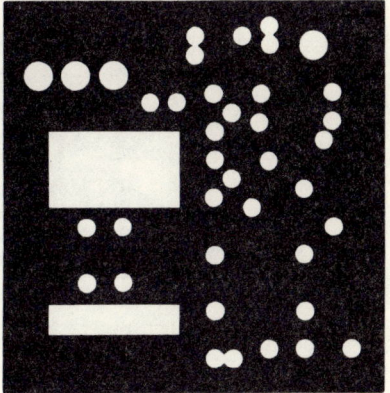

Front　　　　　　　　　　　　Back

Fig. 5-6 Artwork for a double-sided board. In this case the back acts as a shield for the wiring on the front of the board.

Regardless of the process used, you need two masks for the finished board.

Registration
The critical part of making the artwork for a double-sided board is making sure that both parts are in alignment. Alignment of artwork is called *registration*. Registration of both the artwork and the mask is important if the pads on the top of the board are to align with the pads on the bottom of the board.

One method of ensuring registration of the artwork is to use metal pins taped to a flat surface. Holes are punched in the tracing paper to be used for the artwork, as shown in Fig. 5-7. This method gives perfect registration of the artwork even when one of the layers of tracing paper has to be removed. Sometimes it is better to use a Mylar tracing paper because it is stronger and will not tear around the holes. This method of using pins for registering multilayer artwork is also used by the printing industry.

Another method of registering artwork is to use bull's-eyes, as shown in Fig. 5-8. Two or more bull's-eyes are positioned on the bottom sheet of the artwork. Each additional layer of artwork has bull's-eyes placed directly on top of those on the bottom layer. When the layers are in perfect registration, the bull's-eyes appear as one.

Board Exposure
The negatives for double-sided boards are produced like those for single-sided boards, with a few basic differences. Making sure of the registration of the artwork on the circuit board can be done in one of

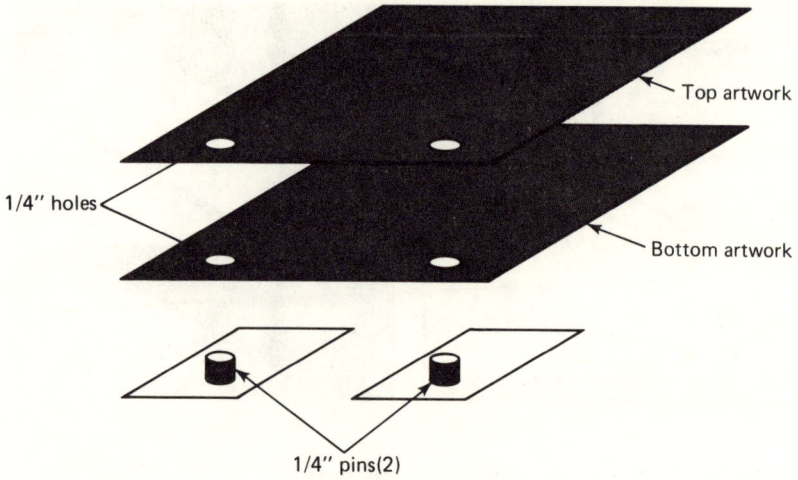

Fig. 5-7 Metal pins are used to aid in keeping alignment tight when preparing the artwork for a double-sided board.

two ways. One is to use the registration pins shown in Fig. 5-9. The double-sided board with its photoresist is sandwiched between the two layers of masking artwork. Each side of the board is exposed to the light source for the correct time. The board is then carefully placed in a developing solution. Be sure that the bottom side of the board is not scratched or handled until it has been developed and the developer has had a chance to harden.

The other way of getting correct registration on both sides is to use masking or transparent tape to hold everything together. Figure 5-10 illustrates this approach. This is the simplest way of achieving good registration. The two masks are joined at one end with tape. The board is then placed between the masks and a second piece of tape is used to hold the package onto the board. Exposure and developing are the same as before.

These steps will produce a double-sided board ready for etching after it has been developed and allowed to dry. *Be sure to remember that*

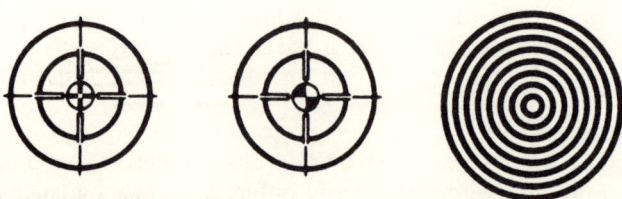

Fig. 5-8 Three forms of bull's-eyes used to align artwork.

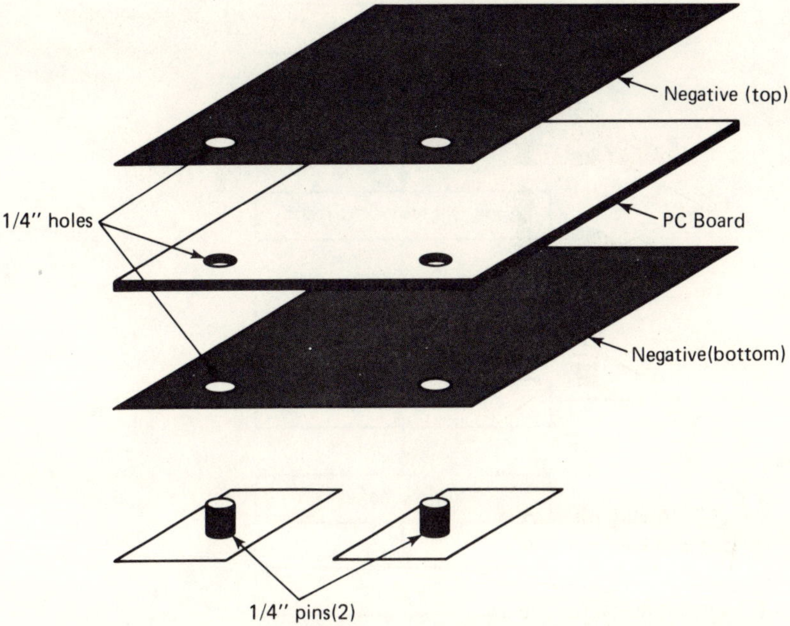

Fig. 5-9 The same metal pins used to prepare artwork are used to align the artwork masks and the double-sided board.

both sides of the board have been exposed. Handle the board by its edges. Be careful not to put finger marks on either side of the board.

PRESENSITIZED BOARDS

Presensitized boards are available from many supply sources. Most boards come in a variety of sizes. Both fiber-glass and phenolic-base boards are available. They usually are packaged in a light resistant black plastic envelope. The package protects the prepared board from

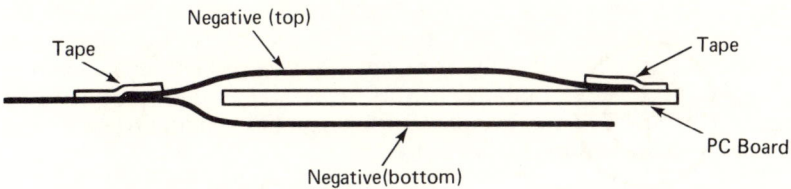

Fig. 5-10 Another system for aligning two negatives and a double-sided board uses tape to hold everything in place.

63

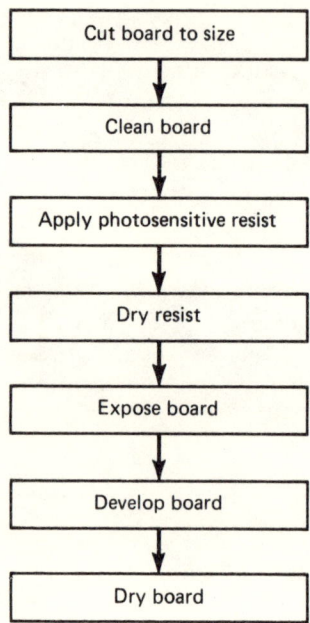

Fig. 5-11 A flowchart illustrating the steps involved in preparing a photosensitized board.

light and should be opened in a semidark room. Use a yellow safety light or work away from direct sunlight. If the package contains more than one board, it should be closed and resealed with tape. Be sure to fold over the opening before taping so that light can't leak in and expose the boards still in the package.

The chemicals used to sensitize the boards evidently have a shelf life, but there seems to be no solid information on this life. The chemicals appear to react with the air and humidity during the storage period. This is not true of all presensitized boards. If you have followed instructions carefully and have poor results, then consider using a new batch of boards or possibly sensitizing your own boards.

This discussion has explained how to make a printed circuit board, starting with sensitizing the board and ending with a board whose surface has a hardened photoresist. The board is now ready for etching. A flowchart showing the steps we have discussed is shown in Fig. 5-11. The steps outlined in this chart are used to produce any quantity of boards. Another process for quantity production of circuit boards is described in the next chapter.

6

Screen-Process Procedures

Producing circuit boards using a photosensitive resist material is relatively expensive. Boards purchased with a photoresist cost about 3 times a nonsensitized board. Hobbyists who plan to produce quantities of one circuit board may wish to use less expensive methods like the screen process.

The screen process uses a resist ink applied through a stencil, or mask, to the surface of the blank circuit board. The stencil is produced and attached to a fine-mesh metal, nylon, polyester, or silk screen. The resist ink is forced through openings in the stencil onto the surface of the blank board. This process produces a positive of the circuit on the copper foil. When dry, the board is ready for etching.

PREPARING THE SCREEN

The frame and screen are illustrated in Fig. 6-1. Most frames used by the hobbyist are made of wood. A piece of screen fabric material is stretched tightly over the frame. Some frames have a groove into which a piece of rope is forced. The rope helps tighten the screen material on the frame. Commercial screens are often made of a fine wire mesh. Special frames are used to hold these screens.

The frame is hinged on one end to the base of the printing board, as shown in Fig. 6-2. It is best to use a hinge with a removable pin. The base of the printing board should be larger than the frame you are using. Equipment required for this process includes a photosensitive film, developer, block-out solution, a rubber squeegee, and the resist paint or ink.

A presensitized film is used to make the screen stencil. A major source of this film material and related screen-process supplies is the

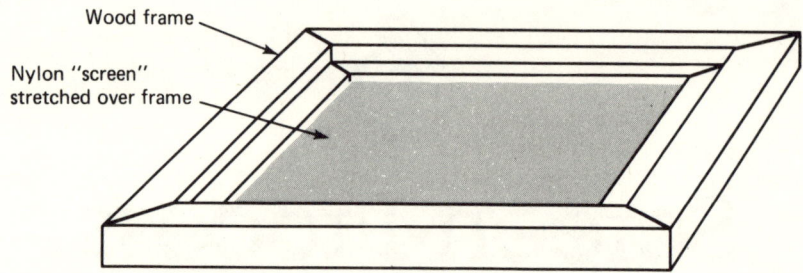

Fig. 6-1 A wooden screen frame. This unit is available from commercial screen-printing suppliers.

Ulano Company. Ulano produces several types of screen stencils. One of the best for use by the hobbyist is their Hi-Fi Green. This film consists of a layer of photosensitive film and a clear plastic backing. This film is soluble in water until it is exposed and developed. After processing, the film is still water-soluble but to a much smaller degree.

The exposure process is very similar to that used to make a photographic negative. A black sheet of paper is placed on the exposure board. A piece of film larger than the circuit board is placed film side down (plastic backing side up) on top of the paper. The artwork is placed on top of the film. They are all held in place with a piece of untinted glass. (Tinted glass would filter the exposure light and affect the quality of the screen stencil.) Exposure is about 8 minutes using a no. 2 photoflood lamp. The lamp should be about 18 in above

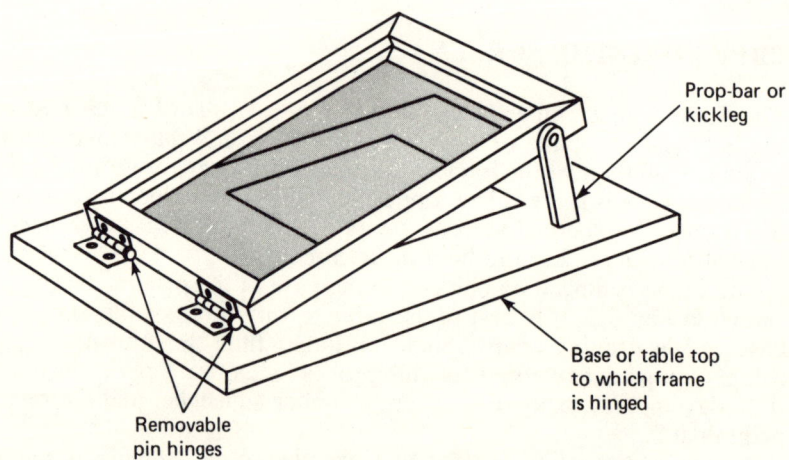

Fig. 6-2 The frame is fastened to a wooden base with pin hinges.

the film and glass. Placing the lamp closer will heat the film and dry it out. This will result in poor adhesion of the film to the screen.

After exposure the film is put in a tray of developing solution for 1½ minutes. Only the developer available from Ulano should be used for this part of the process. It is sold in two packages, identified as Ulano Hi-Fi A and B Developer. The directions state that the temperature of the developing solution should be between 64 and 75°F. Put the film in the tray emulsion side up. Ulano recommends making enough developing solution to cover the film with about ½ in of developer.

The developing solution is not light-sensitive as it is prepared. Once the film is placed in the solution, the solution becomes sensitive to light. At this point the developer should be protected from light and covered. The developer has a maximum life of 24 hours and should not be kept longer than that. Fresh developer is made for additional film processing as required. The tray holding the film and developer must be agitated during the development.

After the required time the film is removed from the tray of developing solution. It is then put in a running bath of warm water (100 to 105°F). Placing the soft side down on the bath may ruin your work. *Be sure to get the emulsion side up.* The purpose of this process is to wash out the film and remove the areas which will permit the ink to be printed onto the blank circuit board.

The washout water should flow over the stencil. This will wash away the unexposed emulsion on the film. Use a gentle water flow, preferably with an aerator nozzle; otherwise you may damage the film stencil. Be sure to wash the entire surface of the film to remove the emulsion completely from the unexposed areas.

This process requires several minutes to complete. To test for a good washout, hold the film over a clean piece of paper. If the color of the water dripping on the paper is clear, washout is complete. If the color of the dripping water is any shade of green, more washout is required.

After the washout is finished, the film has to be cooled. Reduce the washout water temperature to that of the room in which you are working. This will firm the screen stencil. It also helps to keep sharp edges on the lines of the stencil.

Return the film to a flat surface. The glass used to hold the film during the rinse process is convenient, but any solid, flat surface can be used. In this step stencil film is attached to the screen. In order to achieve good adhesion the film stencil must be held on a flat surface and must be raised up higher than the surface you are working on. This is illustrated in Fig. 6-3. Use a hard buildup of about ½ in.

Gently lower the screen and frame onto the film. No force is necessary. The weight of the frame is sufficient to press the screen material into the soft film emulsion.

Place a pad of *unused* newsprint (the kind of paper newspapers use) on top of the screen. Gently wipe over the newsprint with a soft cloth.

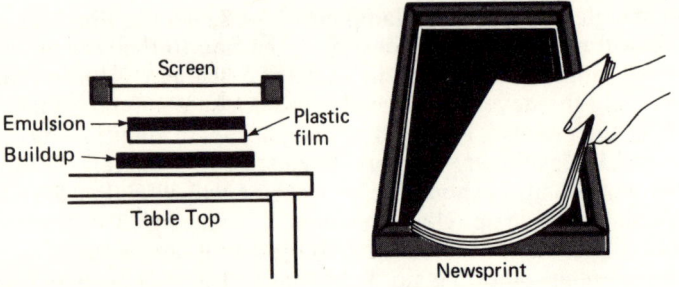

Fig. 6-3 Making the film adhere to the screen. The buildup is required for good adhesion. (*Ulano Companies.*)

This will make the soft emulsion work up into the screen. *Do not use pressure* for this operation. Change the newsprint until it no longer shows any of the green film color. This may require four or five changes. *Do not use printed paper for the pad. Only clean newsprint will work successfully.* Paper towels are not suitable because (1) they do not absorb enough moisture from the screen film, (2) patterned towels often leave patterns in the screen, and (3) towels do not permit good adhesion to the screen.

Once most of the moisture is removed, the screen is set aside to dry. This takes 45 minutes to 1 hour. When the emulsion is dry, the backing material is carefully peeled off the film. The result is a negative mask attached to the screen.

Inspection of the screen under light quickly shows that the fabric is open between the film edge and the frame. These areas must be filled in; otherwise the ink or paint will be forced through the holes in the screen in areas where it is not wanted. There are two ways of correcting this. One is to prepare a paper mask cut to fill in the open area between

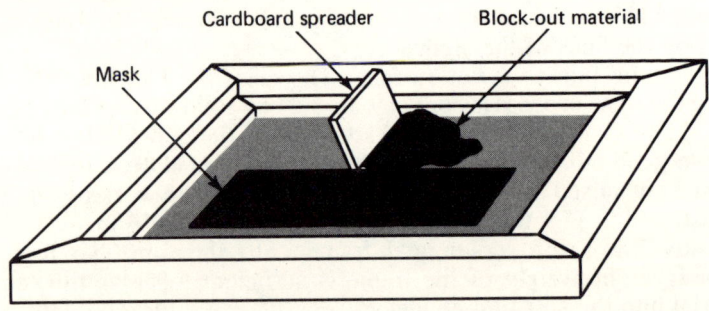

Fig. 6-4 Block-out is used to protect areas of the screen not covered by the mask.

film and frame. It is held in place with masking tape on the bottom side of the screen. The second way is to use a block-out material soluble in either lacquer or water. In this case, since water-soluble film is used, use a water-soluble block-out material. The block-out is spread in the open areas of the screen. Use a stiff piece of cardboard as a spreader. This is illustrated in Fig. 6-4.

Check the screen again for leaks in areas which should be protected. Hold it under a strong light. Touch up with a small paintbrush and block-out material thinned with water. Speed up drying by using a fan to force air against the screen. Keep the screen at least 12 in away from the fan. A flowchart illustrating this process is shown in Fig. 6-5.

PRINTING-BASE PREPARATION

You now have a screen to which a stencil is attached. Inspection shows that all areas to be protected are filled in. Areas which are to appear on the board as conductors or pads are clear. There is no block-out on the screen in these places. The next step is to prepare a printing base.

As explained in Chap. 5, registration refers to correct alignment of material for the printing press. Each piece of paper is held in exactly the same position in order to duplicate the work. This is also true in making duplicate printed-circuit boards. The blank board is placed on the printing frame in exactly the same place each time. Registration guides are used to ensure good registration. Pieces of cardboard or scrap circuit board are taped or tacked in place on the base board (see Fig. 6-6). The registration guides are placed on three sides of the area surrounding the blank board. Sometimes a small piece of double-sided tape is placed in the center to hold the board in place.

The screen frame is held to the printing board base with two hinges. Using a hinge with a removable pin will made cleaning easier.

Now for actually printing the resist on the board. Place a cleaned blank board copper side up in the registration area of the base. Lower the screen and frame so that the screen rests firmly on the copper foil. Pour some of the resist ink or paint on the screen in one of the blocked-out areas. Take a rubber squeegee, preferably the kind used for screen printing, and pull the ink over the negative part of the screen (Fig. 6-7). A piece of stiff cardboard may be substituted for the squeegee. Continue past the negative into a block-out area. One pass of the ink over the negative part of the screen should do if you have used enough ink. Gently lift the screen frame. The board with a resist pattern of the circuit should be seen in all its glory! Carefully lift the board from the registration area on the base. Set it aside to dry.

The purpose of the small piece of double-sided masking tape is to hold the printed board down against the base when the frame is lifted. Large boards will probably stay put because of their weight, but small

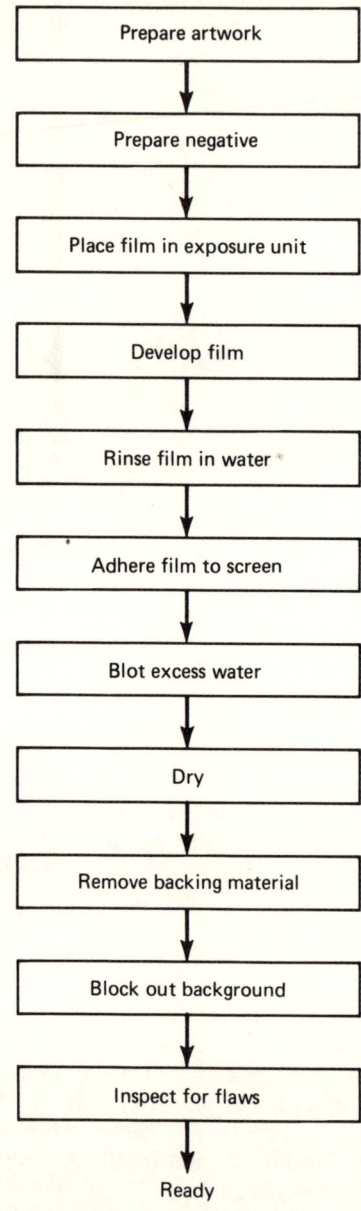

Fig. 6-5 A flowchart showing the steps in making a photographic screen.

Fig. 6-6 Placement of registration guides helps keep alignment consistent for printing.

boards are often held against the back of the screen by the capillary action of the ink. Removing small boards by hand may cause a slight smearing of the wet resist ink. Minor errors are easily corrected by using an artist's knife to remove excess ink or a paintbrush to add missing ink. The board is ready for etching when the ink is dry. Follow the manufacturer's advice for drying time of the ink.

RESISTS AND INKS

There are several different kinds of ink resists available. Some use a water-based ink, and others use a lacquer or oil base. The idea is to use an ink and an emulsion for the film which are of a different base material. This makes it easier to clean up after accidents, and the screen is easier to clean too. If you use a water-soluble film for the mask, choose either the lacquer- or oil-base ink. If you use a lacquer-base film, use either an oil- or water-base resist ink. Since most of the film available for use is water-soluble, let's stick with that and look at each of the other two base materials.

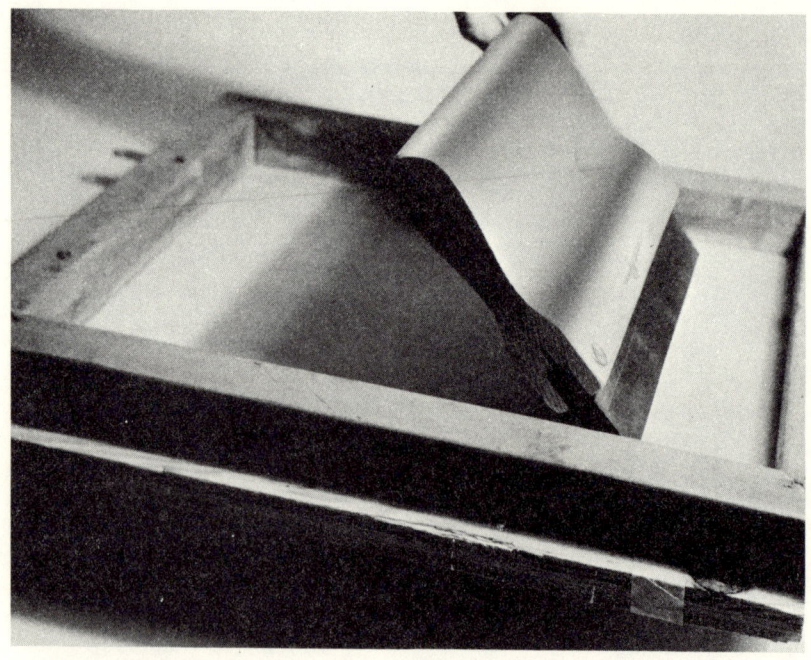

Fig. 6-7 Pulling the ink over the screen with a rubber squeegee.

Oil-base ink is much slower-drying than lacquer-base ink. This means that the clean-up may be delayed with oil-based products. On the other hand, lacquer-based ink dries faster and thus speeds up the overall processing time of the board. Take your pick since it really doesn't make a lot of difference which type of ink you use. Just keep in mind that using noncompatible ink and emulsion materials allows you to wash up the screen and save the stencil. If you use ink and emulsion of the same material, washing one will remove the other. If you have only one production run of the boards and never intend to run that circuit again, and if you can produce all the required boards in one operation before the ink on the screen dries, you can use water-soluble ink and emulsion. While you are making this decision, check the ink resist to be sure that a water-soluble resist ink will not wash off during etching. Better still, don't use a water-base resist ink. You will be much happier in the long run.

CLEANING UP

All the boards are now printed with a resist ink of the proper kind. Each board is laid out, next to its duplicates, and drying. You have decided to save the screen in case you need it again. You are now faced with a big mess. This is called clean-up. Face it! There is no way of doing this job that is fun, clean, or easy. It requires a fairly large work area.

Let's assume that you have selected an oil-base resist paint. This requires a solvent such as paint thinner or terpoline. Start by placing several layers of newspaper between the base and the screen frame. Have a roll of paper towels or other wipe-up material handy. A large waste container lined with a plastic garbage bag is ideal to hold the messy papers when you are done with the clean-up.

Use a spatula, a piece of cardboard, or any other scraper to remove any large blobs of resist ink left on the screen. The rubber squeegee will help collect them. Next, pour some of the solvent onto the screen. Use the wiping towels to mop up the ink and solvent on the screen. Repeat this process until most of the resist ink is removed. You may have to change the newspapers occasionally during clean-up. You'll have to change them often if you have a lot of resist ink to remove.

When most of the ink is gone, you are almost finished. Remove the pins from the hinges, lift the screen frame, and inspect the screen. Look for places where the resist ink has filled in holes in the screen. Use more solvent and clean wiping cloths to clean these areas. When you are satisfied that the board is free of resist ink and ready for reuse, store it away. Check to see that the spatula, the squeegee, and the base are free of resist ink. You have now completed the job.

The next chapter discusses etching to remove the copper foil in areas of the board without resist. It is one of the last steps in processing the board.

7

Etching the Board

All of the work done so far has been to prepare a blank copper board for etching. Using one of the several procedures described in preceding chapters, you have taken a blank board and prepared its surface. This preparation really sets up two surfaces on the board. One will resist, or stop, the etching action. The other will react to the etchant. Enough time and the proper temperature will produce the etched board. The circuit conductors are left on the board after the rest of the copper has been eaten away. This chapter tells how to set up an etching tank and how to etch the board.

ETCHING SOLUTIONS

Theoretically, one of four solutions can be used to etch a board, ammonium persulfate, chromic acid, cupric chloride, and ferric chloride. Other solutions can be used, but they also attack the adhesive bonding material, the base material, or the resist material. The most common etchant used by the hobbyist is ferric chloride ($FeCl_3$). It is the least expensive of the chemicals and the least dangerous. It is readily available at most outlets selling printed-circuit supplies.

Etchants may cause irritation to the eyes or skin. *Always work in a well-ventilated area or room.* Ferric chloride stains the skin orange-brown. It also leaves rusty spots on clothing or floors. Wash any spills immediately with clear water. If spots remain on clothing, a home laundry rust remover available at the grocery store will often remove them without damaging the material.

Ammonium persulfate contains mercuric chloride. This substance is a poison if swallowed. Work carefully and wash your hands in running water if they come in contact with the etchant.

Never use kitchen utensils for holding etching solutions. If you like their size and shape, buy a set for your own use and keep it with the etching supplies. Washing the utensils and returning them to the kitchen is dangerous because they may not be completely clean.

Residue of the etchant solution could come in contact with food and poison someone.

With etching solutions, out of sight is not necessarily out of mind. Both the common etchants love copper plumbing. Remember that they etch the copper foil on the board. They also etch any other copper they come in contact with. If you store the etchants, use a glass or plastic container. Stainless steel can also be used, but is not recommended because stainless steel and brushed aluminum look a lot alike. One will resist the etchant, but etchants love aluminum. These words to the wise also hold true for etching tanks. Be careful of the material used as an etching tank.

When you dump the used etchant, dilute it with large quantities of water to reduce etching action to a minimum. Check with local sewage rules for chemical dumping if you feel that pollution may be a problem.

Etchants are purchased in liquid or dry form. Liquids are premixed and require no further mixing. Dry etchants are mixed with hot water to dissolve the crystals. Ferric chloride crystals are mixed in the ratio of 500 grams $FeCl_3$ to sufficient water to make 1 liter. This quantity will produce about 33 ounces of liquid.

Ammonium persulfate is provided in a plastic bag. When it is mixed with the proper amount of hot water, it gives the correct solution strength for etching. Etching kits containing the chemicals required are available at local outlets.

ETCHING TANKS

Etching is done by tray rocking, a spraying device, or in a tank. Unless one is going to etch in large quantities, either the tray rocking or etching tank will be used. Large, sophisticated etching tanks are available, but special controls and other features make them too expensive for the hobbyist. Most of the printed-circuit boards made by the hobbyist are produced in either a rocking tank or a small etching tank. Let's look at these two processes.

Rocking Tanks

Figure 7-1 illustrates a rocking tank in its simplest form. The tank is a glass baking dish with marbles in the bottom. Etchant is poured into the tank so that it covers the marbles and will also cover a circuit board blank placed on top of the marbles. It should cover the resist. Carefully pick up one end of the etching tray. This causes the marbles, the board, and the etchant to move to the low end. Then gently lower the end of the tray to the table again. The contents should move back in the tray to a point where all is level. This process is repeated until all the areas of the copper board without resist have no copper left on them. This process takes about 30 to 60 minutes, depending upon the freshness of

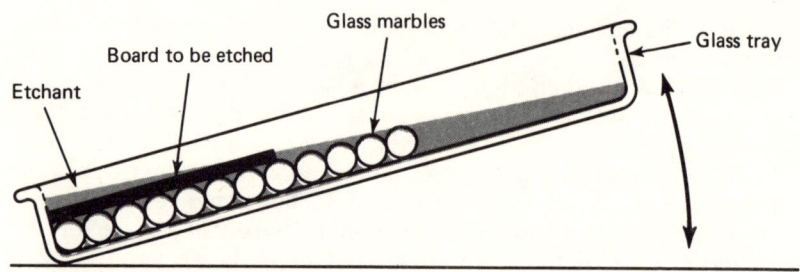

Fig. 7-1 Glass marbles in the bottom of a glass tray make a good etching tank.

the etchant and the size of the board. The same process will work without the marbles if they are not readily available.

The process can be speeded up if the etchant is heated to between 100 and 130°F. *Higher temperatures will produce toxic fumes. Take care not to breathe these fumes.*

The ferric chloride reacts with the unprotected copper areas on the board during etching. The etching action produces a sludge, which settles to the bottom of the etching tank. The formation of this sludge weakens the etching solution. When the etching process slows down considerably, it is time to change the etchant. Sludge also forms in areas where etching occurs. This is why it is necessary to keep the liquid etchant and the board moving in the tank. Rocking the tank allows the areas to be etched to be exposed to fresh etchant and thus speeds up the process.

A "second generation" rocking tank (Fig. 7-2) uses an electric motor and a cam to provide the rocking required for etching. A slow-speed motor (between 1 and 10 revolutions per minute) is adequate for this purpose. The etching tray is held on a hinged board by its own weight. The board is hinged at one end of the tray, and an offset cam is attached to the motor shaft. The offset part of the cam moves the tray and board up and down. This duplicates the motion of hand rocking and allows the etcher to attend to other work while the etching is taking place.

In this version a heat lamp is suspended over the etching tank to heat the etching solution to the desired temperature. Use your imagination if you decide to make your own rocking-tank setup. Just remember that the etchant attacks metal, including aluminum and steel. All parts exposed to the etchant should be glass or plastic. Parts which must be made from aluminum or steel have to be protected from the etchant.

Aerator Tanks

The etchant is kept moving to bring fresh etchant in contact with the board. Once the etchant reacts with the copper, the chemical action reduces the effectiveness of the etchant at that point. Bringing fresh

Fig. 7-2 A homemade etching tank uses a slow-speed motor and a cam for moving the etchant in the tank.

etchant to the surface of the board keeps the etching process going. The rocking tray does this very well.

Another way is to make the etchant move past the board. An easy method is to cause bubbles of air to move across the surface of the board. This is easiest when the board is held vertically in the tank and the air bubbles rise across its surface. See Fig. 7-3. The etching tank used for this purpose is a case from an automobile battery. The case is prepared by removing all the plates and dividers. It is cleaned thoroughly to remove all traces of battery acid.

Go to the pet shop and buy an air pump, an aerator, and some plastic tubing. The salespeople will help you decide the size and shape of aerators that will suit your requirements best.

One or more aerators are cemented to the bottom of the etching tank. Rubber or plastic air hoses connect the aerator to the air pump. Something is needed to suspend the circuit board in the tank. A simple way is to use plastic clothes pins to clip the board against the side of the tank. Be sure to place the copper side of the board out, facing the etching solution and bubbles. Heat the etchant with an aquarium heater or a heat lamp 100 to 130°F to speed up etching. Turn on the aerator. The bubbles should flow past the copper side of the board. Adjust the board if necessary to ensure maximum bubbling action against its surface. Remember that some clothes pins have metal springs. Even if you keep the metal out of the etchant, they do not last long because of the fumes.

Spray Etchers

Still another method of keeping the etchant moving on the surface of the board is to use a spraying system. Spray tanks are used by

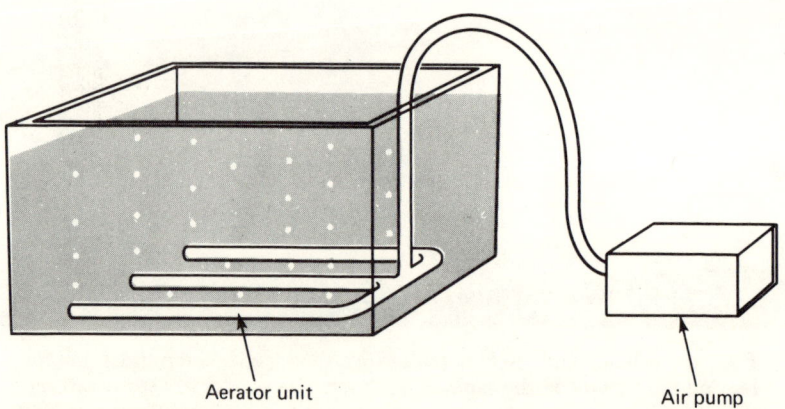

Fig. 7-3   A battery case and an aquarium aerator also make a good etching tank.

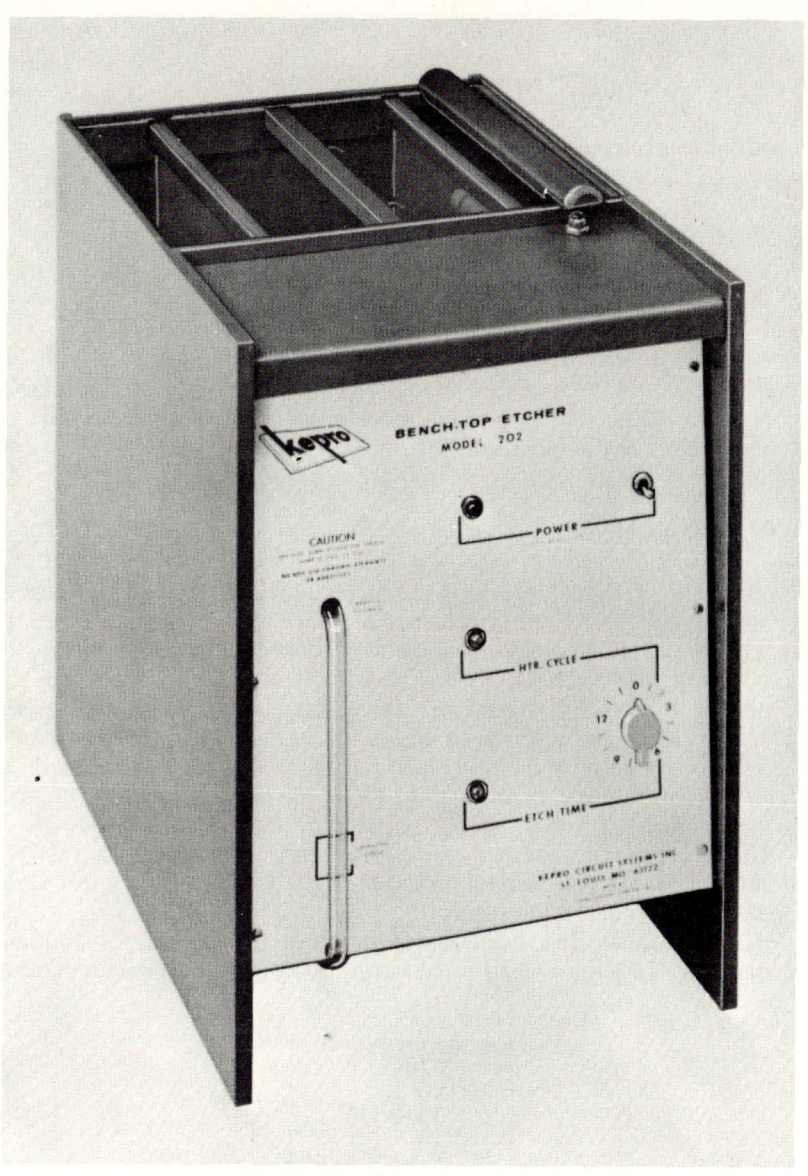

Fig. 7-4 A commercial spray etching tank. This model will spray etchant on both sides of the board at the same time. (*Kepro Circuit Systems, Inc.*)

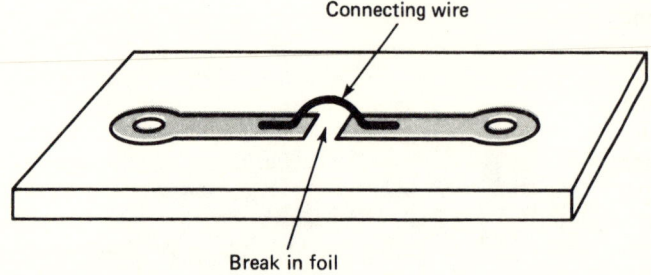

Fig. 7-5 Bridge small breaks in the foil pattern by soldering wires across the break.

commercial board processors, but very few are found in the hands of the hobbyist. A commercial model is shown in Fig. 7-4. This tank contains a heater and a spraying mechanism. Many units also have a timer to control the length of time in the etchant.

AFTER ETCHING

Each of the methods described in this chapter will produce an etched printed circuit board. After the board has been etched, it is inspected to make sure that the circuit is good. Look for hairline breaks in conductive paths. Check for unwanted connections between pads or conductors where the resist may have bled over the desired area. These bridges can be removed with a sharp knife. If there are many, scratch or rough up the surface and put the board back in the etching tank for a few more minutes until these unwanted areas are etched away.

Use a lacquer thinner or resist remover to remove photo resists. Paint thinners can be used on oil-base resists. Use whatever means are necessary to remove the resist.

After the resist has been removed, clean the copper for another inspection. This time small breaks can be bridged with solder. Large

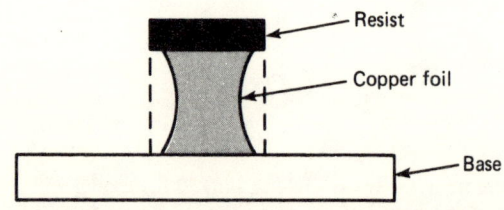

Fig. 7-6 Overetching the board will undercut the foil left on the board and reduce conductor width. This also causes breaks in the conductive paths.

breaks may require a short jumper wire soldered across the break in order to make the repair. Figure 7-5 shows how. If too many breaks appear, something was wrong with either the original resist application or the etching process. If the resist is not firmly attached to the board, a break will occur during etching. Too much time in the etching tank will cause the copper foil to be undercut. This can also produce breaks in the foil pattern, as shown in Fig. 7-6. Avoid this situation by watching the board as etching nears completion.

After the board is inspected and approved, it is ready for the next steps. They include hole drilling, parts mounting, and trimming to final size, discussed in the next chapter.

8

Final Processing

The last act in the production of a printed-circuit board has several scenes. In order, they are: final cutting of the board, drilling holes for mounting parts, cleaning up the holes, mounting the parts, and soldering all leads in place. Each of these is described in this chapter.

CUTTING THE BOARD

The circuit board has to fit into the cabinet. Sometimes the board is laid out to fit a specific commercial size of board. If it has to be cut to final size, the cutting is often done after the board has been etched. Waiting until then saves the effort of cutting a board that contains a bad circuit or is badly etched.

Method 1 There are two methods of cutting a board to final size. One way is with a saw. The board is clamped between two pieces of wood, as shown in Fig. 8-1. The wood is placed very close to the cutting line on the board. The purpose of the wood is to keep the board from twisting and fracturing during cutting. Use a coping saw or hacksaw to cut the board to size. Make all cuts from the foil side of the board to avoid pushing the foil away from the base material. Clean up rough edges with a file while the board is still clamped.

Method 2 The second method uses a paper cutter with a blade attached to a wooden base, the kind photographers use. The board is held in place and the blade of the cutter is firmly lowered against the board. This process may fracture the board or twist on the cutter. Practice a few times with some scrap material before trying to cut a good board. The guillotine cutter shown in Fig. 8-2 can also be used. This is a commercial cutter designed especially to cut the boards. Again the foil side of the board must be on top when cutting in order not to tear the foil.

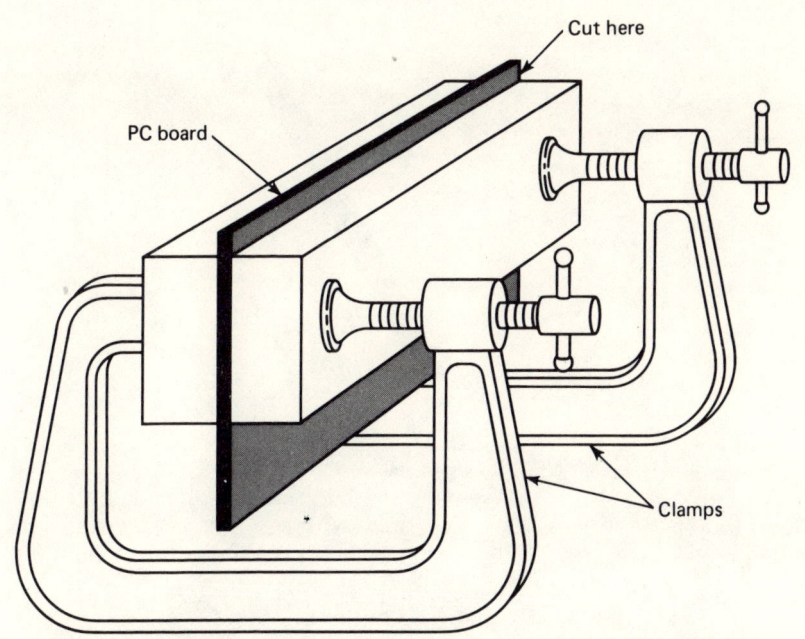

Fig. 8-1 Clamp the circuit board between two pieces of wood when cutting the board with a saw.

DRILLING THE BOARD AND CLEANING THE HOLES

Holes drilled through the board help hold parts in place. The size of the hole, while not critical, is important. Trying to force a component lead through a hole that is too small may damage the board or the component. If the hole is too large, mounting is sloppy and the solder connection may fracture due to vibration. Table 8-1 is a drilling chart showing the size of hole required for mounting common parts. Several sizes on the chart are very close to each other. The purist may wish to purchase several of each size drill, but the practical hobbyist will probably compromise on the drill sizes.

Most holes required for mounting component parts can be drilled with just a few drill sizes. Table 8-2 shows some examples. Drill size numbers increase as the diameter of the drill decreases. Therefore a no. 60 or 65 drill may be used to mount diodes, ¼-W resistors, and small-signal transistors. Components with larger-diameter wires, such as capacitors and 1- and 2-W resistors, may need a no. 55 drill for their mounting holes.

Integrated circuits can also use the smaller size drill. Mounting hardware, such as bolts, requires a variety of holes, depending upon

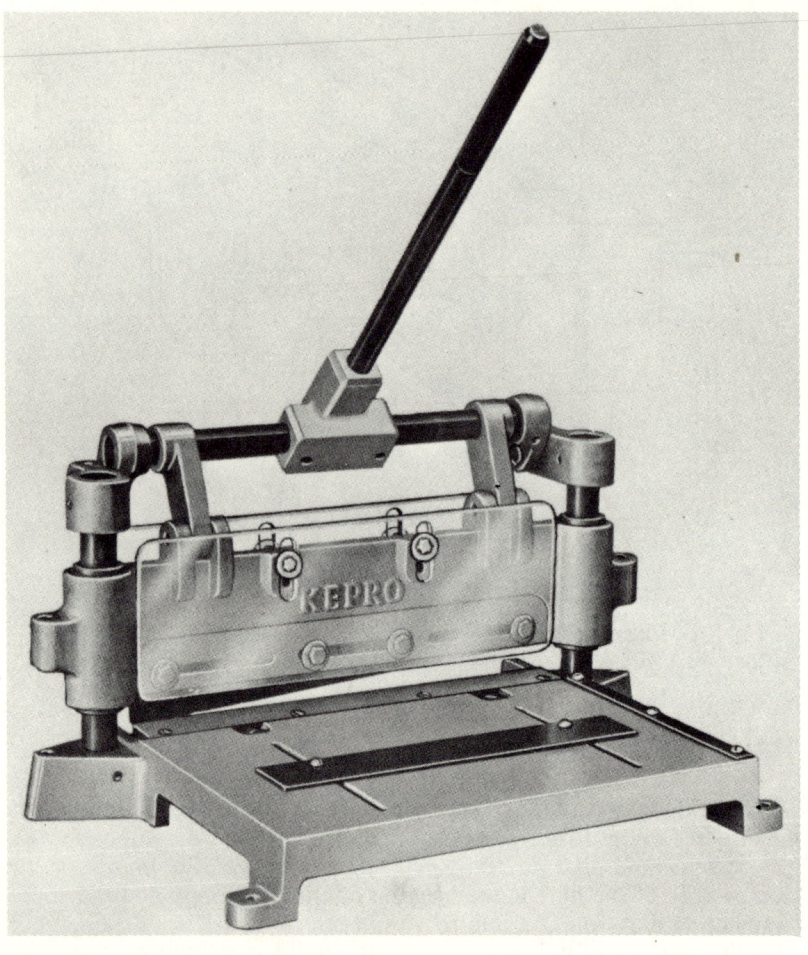

Fig. 8-2 A commercial version of a board cutter. (*Kepro Circuit Systems, Inc.*)

the diameter of the bolt. Table 8-3 shows clearance holes and tapping holes for various bolt sizes. Mounting holes for the power transformer in the variable power supply have a diameter of about 3/16 in. A no. 12 drill will provide the correct mounting hole.

Most circuit-board drilling is done at a high speed. Use a drill press if one is available. Set it at its highest speed. A hand tool will also give the proper drilling speed. Take care not to bend or break the drill bit when using the hand tool.

Place a piece of hard backing under the circuit board when you drill

Table 8-1 Drill Sizes for Electronic Components

Component or device	Lead diameter, in	Drill size, no.	Decimal equivalent of drill, in
⅛-W resistor	0.016	75	0.0210
¼-W resistor	0.019	72	0.0250
½-W resistor	0.027	66	0.0330
1-W resistor	0.041	3/64 in	0.0469
2-W resistor	0.045	55	0.0520
Disc capacitor	0.030	64	0.0360
TO-5 case style	0.019	72	0.0250
TO-18 case style	0.019	72	0.0250
DO-14 case style	0.022	70	0.0280
77-02 plastic power transistor	0.026	67	0.0320
TO-99 (8-pin IC)	0.019	72	0.0250
TO-116 (14-pin DIP)	0.023	69	0.0292

it. This prevents the base material from breaking as the drill cuts through the board. A break around the back of the hole may mean that the component cannot get enough support to hold it in place.

Even the best of drills leave a burr on the metal surface. Burrs must be removed before the parts are mounted on the board. The easiest way to remove this burr is to use a larger drill bit than used for the hole. Hold the bit between your fingers, press it against the hole, and gently twist. Figure 8-3 shows how. Normally all burrs can be removed in this way.

Table 8-2 Only a Few Drills Are Needed to Drill Holes for a Large Variety of Components

Component	Drill size, no.
⅛-W resistor	70
¼-W resistor	70
½-W resistor	65
1-W resistor	55
2-W resistor	55
Disc capacitor	65
8-pin IC	70
16-pin IC	70
Power transistor	65
Small-signal transistor	70

Table 8-3 Drilling Chart for Hole Tapping and Bolt Clearance

	Drill size, no.	
Screw size	Tap hole	Clearance hole
2-56		43
4-40	42	33
6-32	35	28
8-32	29	18
10-32	21	10

SOLDERING

Anyone who is able to lay out and develop printed circuit boards should have a good understanding of solder and flux. Most of the Do's and Don'ts are doubtless familiar, but a few hints will be helpful if you have not done much work with circuit boards.

One trouble that occurs when a circuit board is soldered is that the foil lifts because of excess heat. This can be avoided if you use the technique shown in Fig. 8-4. Start by applying a small amount of solder to the tip of the iron. This will help provide rapid heat transfer from the tip to the work. Next, place the tip of the iron on both the pad and the component lead (Fig. 8-4*b*). Both must be heated to get proper solder

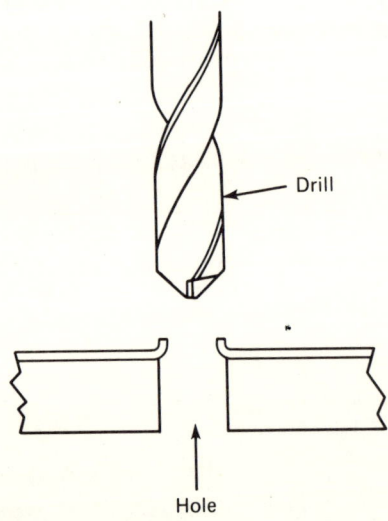

Fig. 8-3 Remove the burr left by the drill before mounting any parts. A larger bit will do this easily.

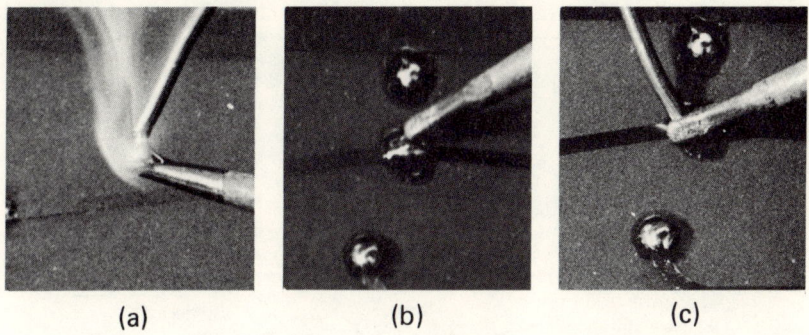

Fig. 8-4 Technique for soldering on a circuit board: (*a*) apply solder to tip of iron, (*b*) position iron to heat both foil and lead, and (*c*) apply a small amount of solder to complete the job.

action. After the joint has heated, usually 1 to 2 seconds, more solder is applied to the connection. If the connection is heated properly, the solder will flow around the lead and the pad (Fig. 8-4*c*). Do not use large quantities of solder. Only a small amount is required for a good connection.

Some of the problems caused by improper soldering techniques are illustrated in Fig. 8-5. It shows the result of not heating the lead before applying the solder to the joint (Fig. 8-5*a*) and of not heating the copper foil during soldering (Fig. 8-5*b*). The illustration also shows a proper solder joint (Fig. 8-5*c*). Inspect your work to be certain you have this type of solder joint.

ASSEMBLING THE BOARD

Only a few of us are very good at standing on our heads. Even fewer have the knack of fitting large fingers in small spaces. There is a procedure for mounting parts that will reduce the frustration of trying to mount a very small part between larger parts on the board.

1. It is easier to work if you have a "third hand". One of these adjustable holders for circuit boards is shown in Fig. 8-6. Several models of this device are available at the parts stores.

2. Mount the parts which are closest to the surface of the board first. They will include diodes and resistors. Turn the board over against the surface of the workbench and solder these leads.

3. Small capacitors, transistors, and integrated circuits will follow. Again, solder these in place. Be sure to have clean leads on the

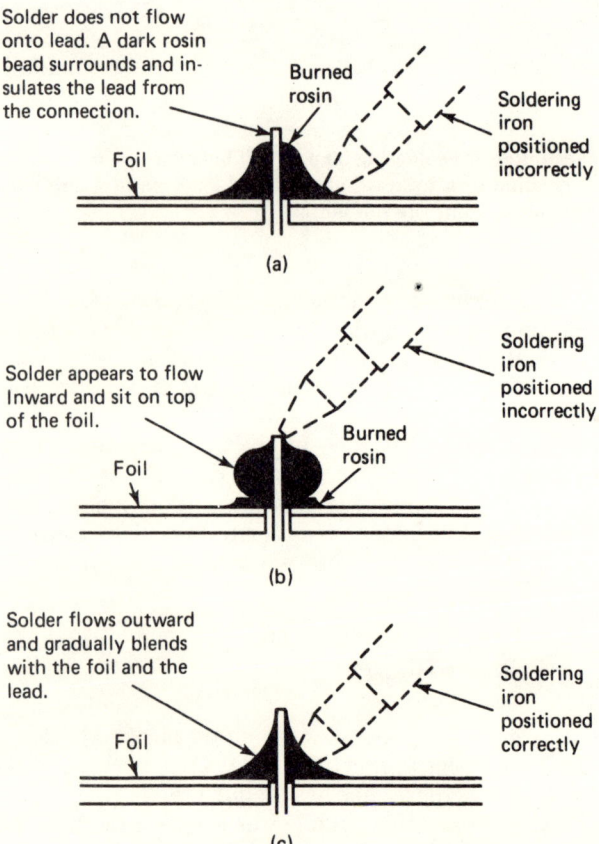

Fig. 8-5 (*a*) **What happens when you don't heat the lead before applying solder.** (*b*) **What happens when the copper foil is not heated.** (*c*) **The correct way to solder on a circuit board.**

Fig. 8-6 Using a "third hand" to hold the board while working on it solves a lot of problems.

Fig. 8-7 The completed circuit board. All on-board parts have been mounted and soldered in place.

components or the solder will not stick to the lead wire. Progress from the smallest to the larger parts in each case.

4. The largest part in our project, the power supply (Chap. 9), is the transformer. It is connected to the board last.
5. Include all wires used to connect off-board components to the board. Clip the excess leads from the parts on the board. Inspect the board to be sure that all solder connections are good.
6. Connect the leads from the board to the proper terminal of the off-board part.

You should now be ready to test the circuit. The board is complete and ready to be mounted in the cabinet. The circuit used for this project and construction suggestions are described in the next chapter. The completed board with all parts connected is shown in Fig. 8-7.

9

Project Description and Construction

One of the most necessary items on the workbench other than testing equipment is a source of dc power. The hobbyist often needs a variety of voltage levels for various projects. The variable dc power supply described in this chapter meets most requirements of the hobbyist. Parts are readily available from a variety of sources (local parts houses and some of the mail-order sources listed in this book or in advertisements in hobby publications). This power supply will serve many of the needs of the home workshop.

CIRCUIT DESCRIPTION

Figure 9-1 is a block diagram of the regulated variable power supply. The schematic diagram for this device is shown in Fig. 9-2. The input is 120 V ac single phase. The ground is connected to the cabinet of the device. A transformer is used to step down the source voltage to 25 V with a capacity of 1.2 A. It also isolates the circuit from the ac line. A fuse rated at 0.250 A is wired between the ac input and the primary of the transformer. Switch S_1 turns the power supply on and off. The switch is physically connected to the variable voltage control R_3. Thus the operator always starts at zero output volts. This will minimize or even eliminate the chance of applying the wrong voltage to a unit under test. Having to start at 0 V reduces the tendency to forget to check the voltage level when first turning on the unit.

The 25 V is rectified by the rectifier network. The network used for this unit is an encapsulated unit. It contains four diodes in one plastic assembly. If a unit of this type is not available, use individual diodes. Each one should be rated at 100 PIV and 2 A. Extra pads will be required on the circuit board if discrete diodes are used. The circuit common floats. It is not connected to the case of the power supply.

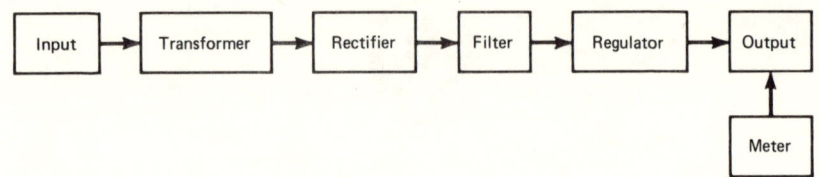

Fig. 9-1 Block diagram of the variable regulated DC power supply.

The output of the rectifier is an unfiltered full-wave dc voltage. This voltage is filtered by capacitor C_1. The filtered voltage is about 35 V dc at this point. It is fed to the input of the regulator chip IC_1. The regulator is a LM317-T. This is a variable regulator made in two different case configurations. The T model, which was readily available, was used in this circuit. If you have access to the K version, the circuit board may require some slight changes. Check this detail before laying the circuit out. Both the T and K versions are shown in Fig. 9-3.

The output of the voltage regulator is controlled by a voltage-dividing network consisting of resistor R_1 and the variable control R_3. Varying the resistance of R_3 changes the voltage present at pin 1 on the LM317-IC. The output voltage of the regulator increases as the input voltage increases. An input voltage drop will produce a reduction in the level of the output voltage in the same manner. Capacitor C_2 is used to improve the performance of the regulator circuit.

The metering circuit consists of a variable resistor R_2 and a meter. The meter used in this supply has a basic 0- to 1-mA movement. It is calibrated from 0 to 250. All readings on the supply are divided by 10 to obtain the correct voltage. Meters with other scales are available from the various sources listed. Select the one that best meets your needs. The meter is wired in series with the variable resistance R_2. This resistance acts as a multiplier resistor to extend the basic meter range in this circuit. The meter circuit is wired in parallel with the output terminals of the regulator. This monitors the output voltage.

CONSTRUCTION TECHNIQUES

The circuit board used for this power supply is shown in Fig. 9-4, along with parts layout for the board. Use the techniques described in Chap. 8 to mount the parts. The heat sink for the regulator is coated with a heat-sink compound during mounting.

Provision for mounting the board was made during the board layout. The two holes for mounting the power transformer are also used to help support the board in the cabinet. Two additional holes are drilled in the opposite end of the board for mounting screws. Spacers made of plastic

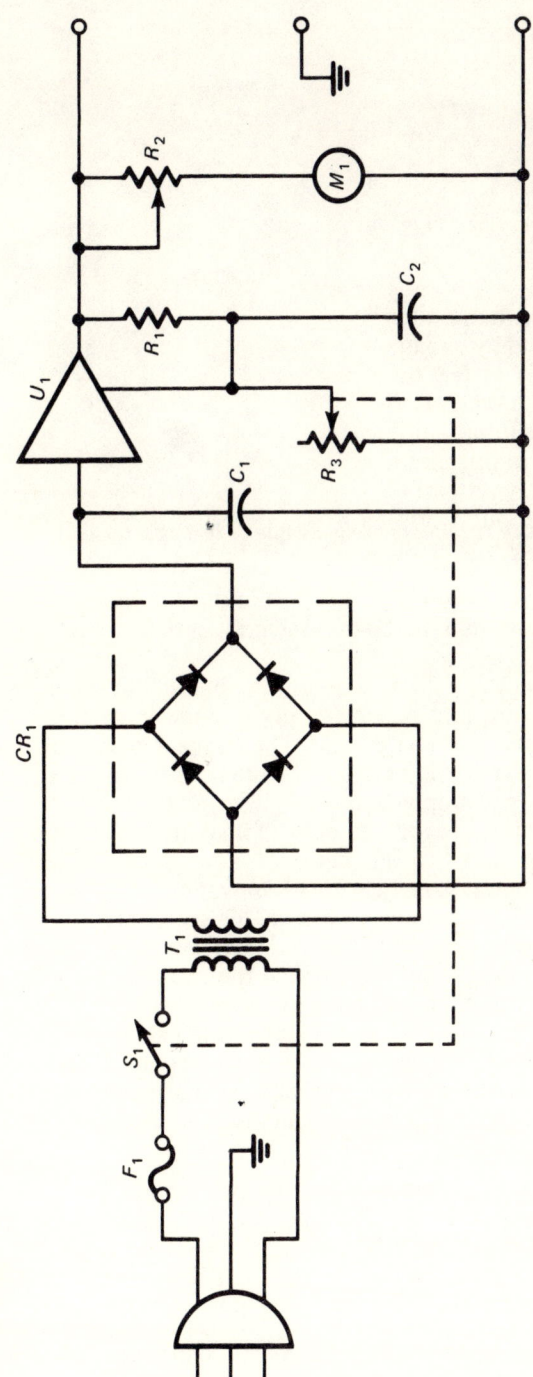

Fig. 9-2 Schematic diagram of the power supply.

Fig. 9-3 Voltage regulators are available in two case forms.

tubing are used to hold the board assembly in the cabinet, as shown in Fig. 9-5.

The interconnecting wires between the board and the panel components are made up into a cable. This is done to facilitate repair or replacement of parts at a later date. The completed assembly is shown in Fig. 9-6. The front and rear panel components are mounted in their respective places. The placement of these parts is not critical. Make your own decisions based on ease of making connections, turning knobs, and reading the meter scale.

If you have bought a meter with a removable plastic cover, you may wish to do the dial scale over. Most meter scales are held in place by two small screws. Remove them and carefully slide the scale away from the meter. Be careful not to bend the dial pointer. (The pointers are difficult to straighten and easy to break.)

The numbers printed on the face of the dial can be removed with an eraser or solvent. Touch up the dial with white paint if necessary. New numbers can be painted on the dial or rub-on artwork numbers can be used. A little care in this operation will produce some very professional results.

FINAL TESTS AND CALIBRATION

Check all the wiring and solder connections. Look for wires or components which might accidentally be touching other parts and separate these. Be sure that all wires and connections are soldered to

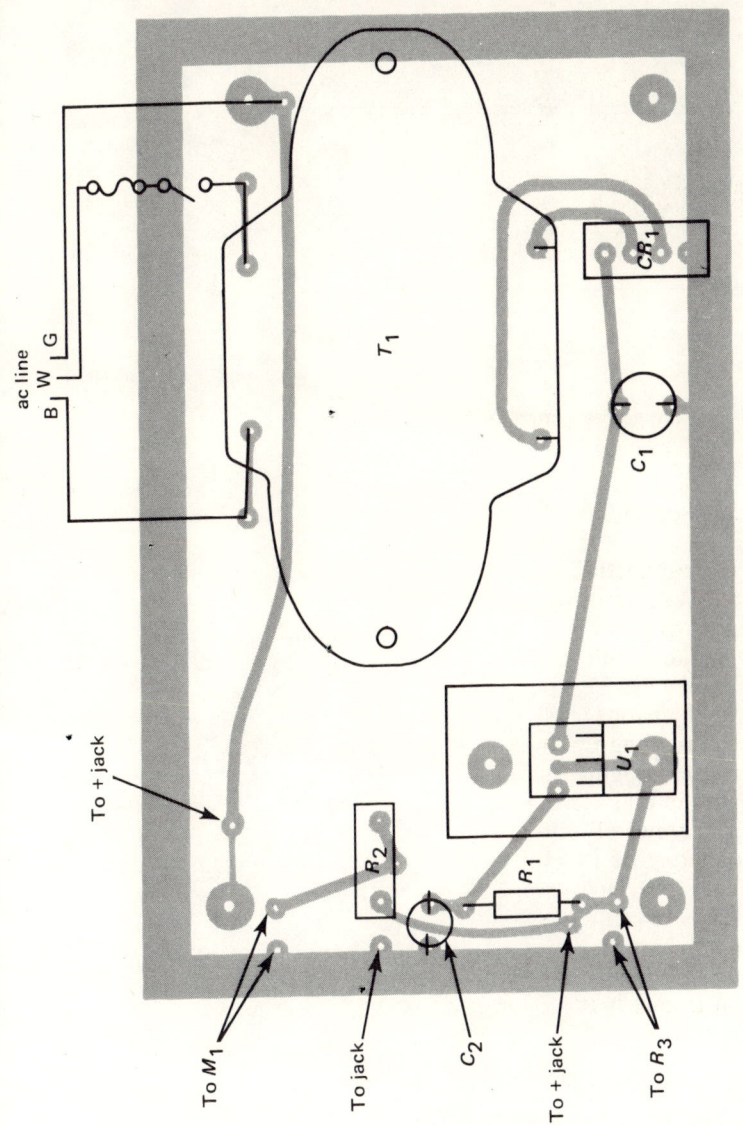

Fig. 9-4 The complete printed-circuit board.

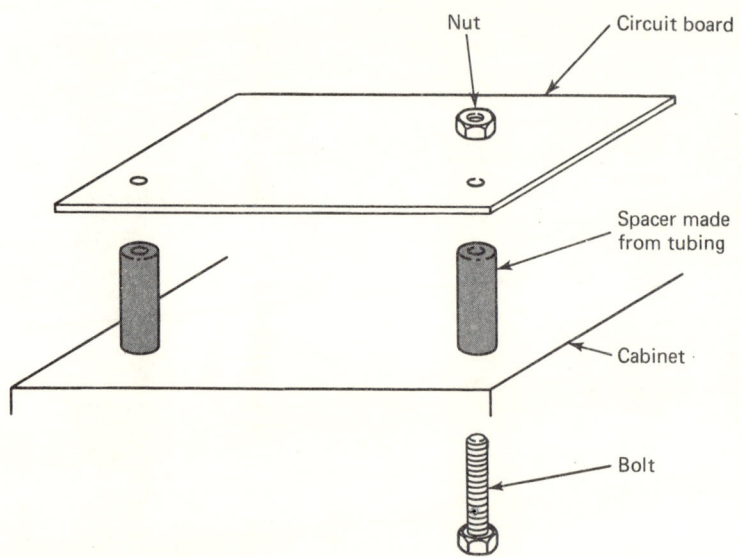

Fig. 9-5 Spacers made of plastic tubing are used to hold the board in place in the cabinet.

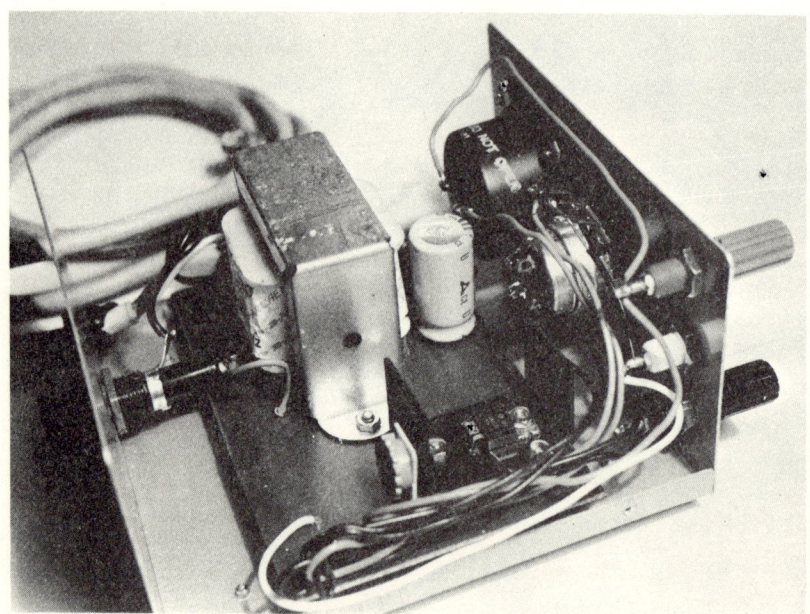

Fig. 9-6 The completed power-supply circuit board and off-board components.

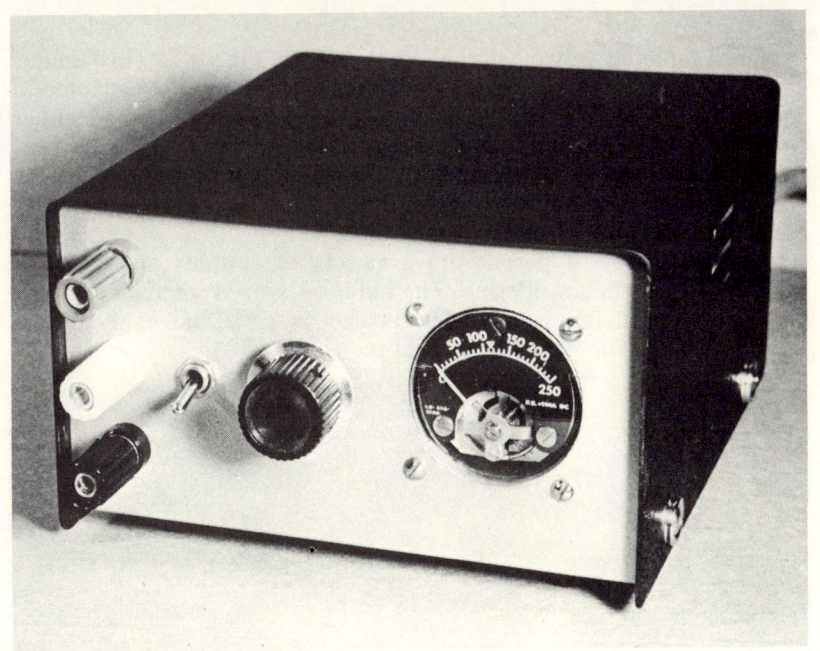

Fig. 9-7 The complete packaged power supply.

their respective connections. Turn the variable voltage control to the "ac off" position.

Plug the unit into an ac outlet. There should be no smoke or other signs of trouble. If all appears well, you are ready for the next major step.

Turn the meter calibration control R_2 to its maximum resistance position. Connect a voltmeter to the output terminals of the power supply. Set the voltmeter range to a minimum of 0 to 25 V dc. Turn on the power supply. If no smoke is observed, you are ready to calibrate the panel meter. Advance the variable voltage control R_3 to a point that shows 25 V on the voltmeter connected to the output terminals. Adjust R_2 so that the meter on the panel also indicates 25 V. Vary the control R_3 and check both meters. They should be moving in step with each other. Both should show the same voltage at any setting of the variable voltage control R_3. There may be slight differences. Do not worry about minor variations in readings.

To check the regulation of the power supply you must have a load connected to the output terminals. A resistor rated at 10 Ω and 15 W will handle this. Three no. 47 pilot lamps wired in parallel will do the same thing. Observe the meter as the load is connected and disconnected from the output terminals. There should be no change in output voltage as this is done.

The completed power supply is shown in Fig. 9-7. This unit is complete with its cover. The basic circuit can be varied or adapted in many ways. These include adding a switch in order to read either current or voltage at the output, using an LED "on" indicator, enlarging the cabinet and building two supplies in it (and thus having a dual output supply), or even adding pass transistors to the circuit to handle larger current requirements.

The availability of parts from a variety of sources makes this a project well worth considering. The hobbyist's need for such a supply makes it a good investment. The project is easy and enjoyable. It should be satisfying to all who build it.

Appendix A

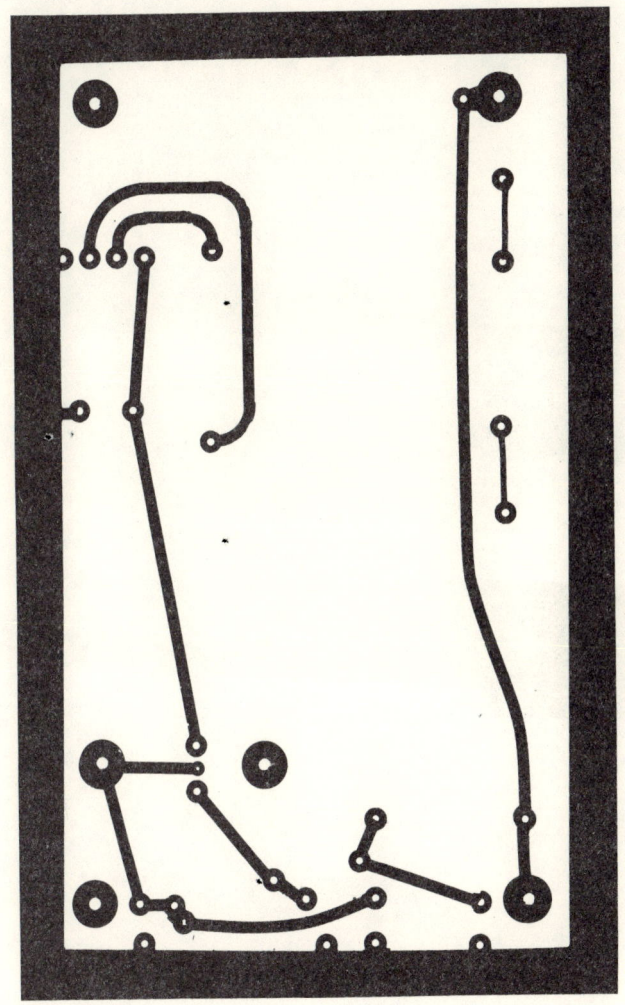

Appendix B

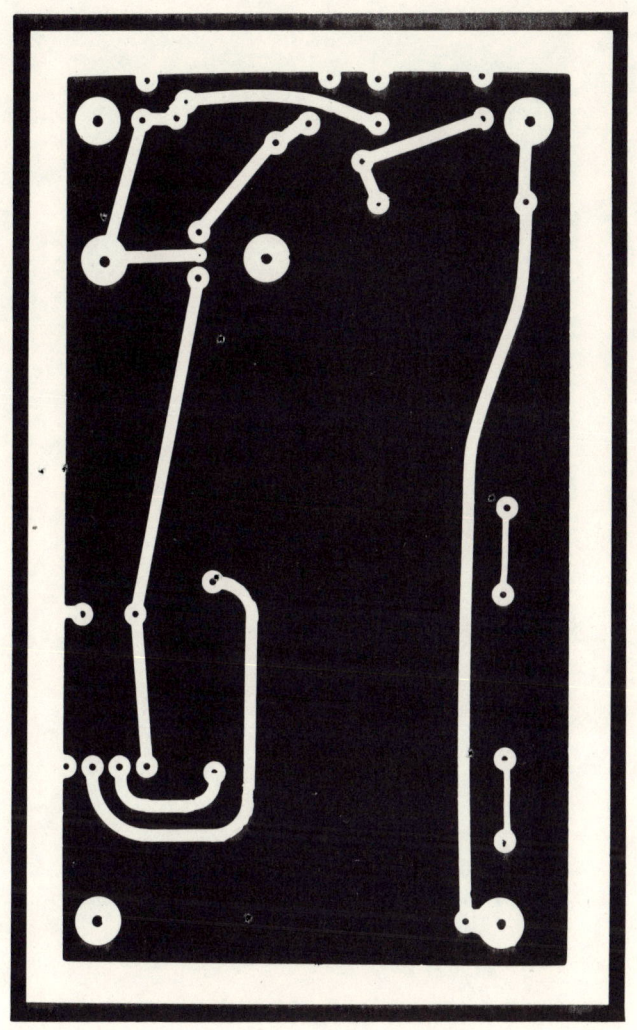

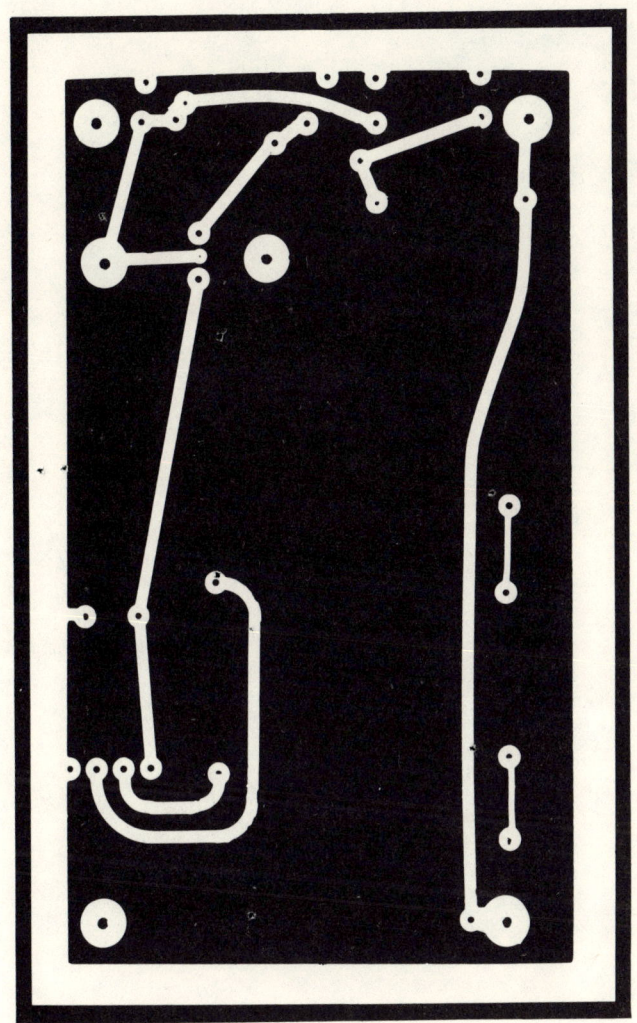

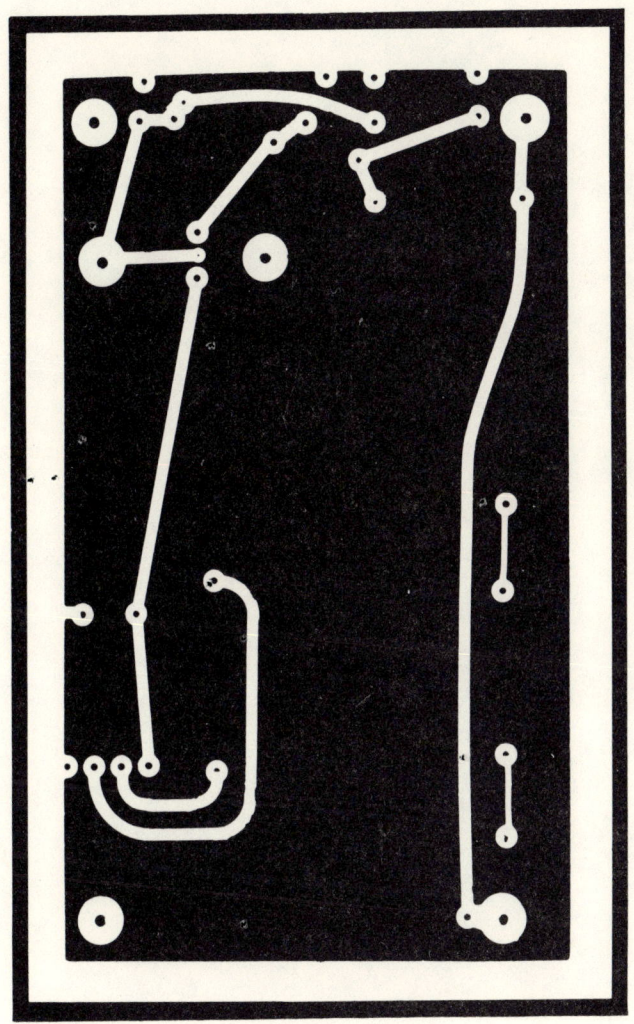

Appendix C

To list every company selling electronic components and supplies for printed-circuit-board production would need a large book. The following list is limited to major sources of supply available at the time of writing. Local telephone directories and the advertising sections of magazines related to electronics will provide additional sources of supply.

The list of sources is arranged in alphabetical order. The columns to the right of the company name and address indicate the types of supplies available from that company. Check for local sources where indicated.

Source	Artwork	Boards	Chemicals	Equipment	Photographic film
Bishop Graphics P.O. Box 5007 Westlake Village CA 91359 (check local drafting-supply sources)	x				x
Circolex Scientific Co. Box 198 Marcy, NY 13403	x	x	x		x
Excel Circuits Co. 4412 Fernlee Royal Oak MI 48073	x	x	x		x
Kepro Circuit Systems 3630 Scarlet Oak St. Louis MO 63122 (check local electronic parts distributors)		x		x	x
Lafayette Radio (local outlets)	x	x	x		
Newark Electronics (local outlets)	x	x	x	x	x
Olsen Radio (local outlets)	x	x	x		

113

Source	Artwork	Boards	Chemicals	Equipment	Photographic film
Printed Circuit Products Co. P.O. Box 4034 Helena MT 59601					x
Radio Shack Corp. (local outlets)	x	x	x		
Rainbow Industries P.O. Box 2366 Indianapolis IN 46206	x	x			
3M Company Industrial Graphics Division 3M Center St. Paul, MN 55101 (check local drafting supply sources)				x	x
Trumbull Co. 833 Balra Dr. El Cerrito, CA 94530	x	x	x		
Ulano 610 Dean St Brooklyn, NY 11238 (check local drafting supply sources)					x

Index

Aerator tank, etching, 76, 77
Ammonium persulfate, 74
Ampere (A), 11, 14
Artwork:
   double-sided boards, 60–61
   guidelines, 22, 25–29
   layout of: dolls used in, 19
      general discussion of, 16–18
      grid paper used for, 19
      guidelines for, 22, 25–29
      methods of, 18–22
      procedure, 29–33
      schematic diagram of circuit, 14–15
      (*See also*, Layout)
   mask making, 51–57
      Circolex process, 52–53
      copy-camera process, 57
      Excel Circuits process, 53–54
      LifTact process, 53
      PCP Transfer Film process, 53
      3M Color-Key process, 54–57
   size of, 17–18, 51
   templates, 21, 23

Base material, 11–12
   considerations in proper selection, 13
   fiber-glass, 12
   frequency ranges for, 12
   phenolic, 11
Board(s):
   assembling the, 87, 90
   bases (*see* Base material)
   checking the work, 35, 37
   choosing the, 16
   classification of, 8–9
   cleaning the, 39
   conductive paths, 34
   construction of, 3, 92, 94
   cost factors, 12–13
   cutting the, 82
   double-sided (*see* Double-sided boards)
   drilling the, 83–85
   etched areas, 35
   etching the (*see* Etching)
   preparation of, 3–5
      for layout, 39–41
   presensitized, 63–64
   selecting proper material for, 13
   soldering on, 86–87

Cabinet, spacers in, 96
Calibration of panel meter, 97
Capacitors, 16, 83
   disc, 21, 85
   radial-lead electrolytic, 21
Chemicals, etching, 74–75
Chromic acid, 74
Circolex photomask process, 52–53
Components, off-board, 25–26, 29
Conductive paths, 27, 34
Conductor paths, spacing between, 17, 18
Copper foil, 9–11, 43, 44
Copy camera, 17–18
Copy-camera photomask process, 57
Cost factors, circuit boards, 12–13
Cupric chloride, 74
Current rating, 11
Cutting, board, 82

Diodes, 32, 83, 91
   power rectifier, 21
   signal, 21
Direct layout procedures (*see* Layout, direct procedures)
Direct resist processes (*see* Layout, direct procedures, direct resist materials)
Dolls in board layout, 19–22, 29, 32–35

115

Double-sided boards, 8–9, 33
 artwork, 60–61
 exposure, 61–63
 registration, 61
Drilling, board, 83–85

Etchants, 44, 74–75
Etching:
 chemicals for (etchants), 74–75
 clean up after, 80–81
 overetching, 80–81
 safety measures, 74
 solutions for, 74–75
 tanks for, 75–80
  aerator, 76, 78
  rocking, 75–76
  spray, 78–80
 temperature of etchant, 76
Excel Circuits photomask process, 53–54
Exposure, double-sided boards, 61–63

Ferric chloride ($FeCl_3$), 74
Fiber-glass boards, 12
Foil, copper, 9–11, 43, 44
 manufacturing, 9
 weight rating, 10

Grid paper, artwork, 18–19, 33

Hi-Fi Green, screen stencil, 66

Ink resists, 45, 47, 71–73

Jumper wires, 28

Layout:
 artwork (see Artwork, layout of)
 direct procedures, 39–48
  board preparation, 39–41
  direct resist materials: household products used as, 47
  pen-and-ink process, 45–46
  rub-on process, 44–45
  rubber-stamp process, 46–47
  tape-and-dot process, 41–44
 guidelines for, 22, 25–29
  common connections, 26
  conductor routing, 27–28
  individual pads, 26–27
  jumper wires, 28
  line direction, 26
  line spacing, 27
  off-board components, 25–26
  schematic diagrams, use of, 25
  separation of inputs and outputs, 26
 methods of, 18–22
 photographic procedures, 49–64
  photomask process, 51–57
   Circolex, 52–53
   copy camera, 57
   Excel Circuits, 53–54
   LifTact, 53
   PCP Transfer Film, 53
   3M Color-Key, 54–57
  photoresists (see Photoresist)
  screen-process procedures (see screen process)
LifTact photomask process, 53

Mask making, 51–57
 Circolex process, 52–53
 copy-camera process, 57
 Excel Circuits process, 53–54
 LifTact process, 53
 PCP Transfer Film process, 53
 3M Color-Key process, 54–57
Mounting pads, 26–27
 integrated-circuit, 46

Negatives, photographic, 50

Off-board components, 25–26, 29
Overetching, 80–81

Pads, mounting, 26–27
Parts layout, 95

PCP Transfer Film photomask process, 53
Pen-and-ink process, direct resist, 45–46
Phenolic boards, 11
Photographic layout procedures (*see* Layout, photographic procedures)
Photographic mask, 51
Photographic negative, 50
Photographic positive, 50
Photomask processes (*see* Layout, photographic procedures, photomask process)
Photoresist:
   application of resist, 58–59
   development of board, 59–60
   double-sided boards, 60–63
   exposure of board, 59–60
   preparation of board, 57–58
Positive, photographic, 50
Potentiometer, 16
Power supply (*see* Variable power supply)
Presensitized boards, photographic layout, 63–64

Rectifier, 16
Registration, double-sided boards, 61
Resist ink, 45, 47, 71–73
Resist materials, 3–4
   developing, 60
   direct (*see* Layout, direct procedures, direct resist materials)
   photo (*see* Photoresist)
Resistors, 16, 21, 32, 83, 85
Rocking tank, etching, 75–76
Rub-on process, direct resist, 44–45
Rubber-stamp process, direct resist, 46–47

Schematic, power supply, 15, 93
Screen process:
   cleaning up, 72–73
   ink resists, 45, 47, 71–72
   preparation of screen, 65–69
      exposure process, 66–69
      frame, 65–66
      stencil, 66
   printing-base preparation, 69–71
Size of artwork, 17–18, 51
Soldered connections, testing, 94
Soldering, board, 86–87
Solutions, etching, 74–75
Spacers in cabinet, 96
Spacing between conductor paths, 17, 18
Spray tank, etching, 78–80
Substrate, board, 3
Supplies, sources of, 104–105

Tanks, etching, 75–80
Tape-and-dot process, direct resist, 41–44
Temperature, etchant, 76
Templates, 21, 23
3M Color-Key photomask process, 54–56
Transformer, 16, 91
Transistors, small-signal, 21, 83, 85

Undercutting, 84

Variable power supply, 15
   board negative, 101, 103, 105
   board positive, 99
   circuit description, 91–92
   construction techniques, 92, 94
   parts list, 16, 29
   schematic, 15, 93
   testing, 94, 97–98
Voltage regulator, 92
Volts (V), 14, 17

Wires, soldered, checking, 94, 97